D1556894

CONSTANTINE SAMUEL RAFINESQUE

CONSTANTINE SAMUEL RAFINESQUE

A Voice in the American Wilderness

LEONARD WARREN

THE UNIVERSITY PRESS OF KENTUCKY

Publication of this volume was made possible in part
by a grant from the National Endowment for the Humanities.

Scholarly publisher for the Commonwealth,
serving Bellarmine University, Berea College, Centre
College of Kentucky, Eastern Kentucky University,
The Filson Historical Society, Georgetown College,
Kentucky Historical Society, Kentucky State University,
Morehead State University, Murray State University,
Northern Kentucky University, Transylvania University,
University of Kentucky, University of Louisville,
and Western Kentucky University.

Editorial and Sales Offices: The University Press of Kentucky
663 South Limestone Street, Lexington, Kentucky 40508–4008
www.kentuckypress.com

Frontispiece: Miniature portrait of Rafinesque
in enamel, possibly by William Birch.
Transylvania University Library, Lexington, Kentucky

08 07 06 05 04 1 2 3 4 5

Library of Congress Cataloging-in-Publication Data
Warren, Leonard, 1924–
Constantine Samuel Rafinesque : a voice in the
American wilderness / Leonard Warren.
p. cm.
Includes bibliographical references and index.
ISBN 0-8131-2316-X (alk. paper)
1. Rafinesque, C.S. (Constantine Samuel), 1783–1840.
2. Naturalists—United States—Biography. I. Title.
QH31.R13W37 2003
508'.092—dc22
2003024567

This book is printed on acid-free recycled paper meeting
the requirements of the American National Standard
for Permanence in Paper for Printed Library Materials.

Manufactured in the United States of America.

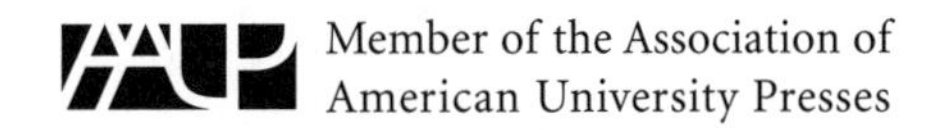

"To see, to know, to publish, became with him a ruling passion."

Fitzpatrick

For those who have shaped me—
Charles H. Best, Zelma B. Miller,
John M. Buchanan, Herbert Tabor,
Seymour S. Cohen, Hilary Koprowski.

CONTENTS

ABBREVIATIONS

ANS	Academy of Natural Sciences of Philadelphia
APS	American Philosophical Society
BPL	Boston Public Library
CP	College of Physicians of Philadelphia
HSP	Historical Society of Pennsylvania
LC	Library Company of Philadelphia
VP	Van Pelt Library of the University of Pennsylvania
WFI	Wagner Free Institute of Science of Philadelphia

CHRONOLOGY

1783	Born in Constantinople, October 22; brought to Marseille, where he was reared.
1792	Family flees to Livorno (Leghorn), Italy, to escape the French Revolution.
1793	Father, a merchant trader, dies of Yellow Fever in an epidemic in Philadelphia.
1797	Family returns to Marseille. Irregular education.
1802	Arrives in Philadelphia. Works in a counting house and begins his botanical expeditions on foot. Advocates the new *Natural* system of classification, which gains slow acceptance in America. Unable to find a position in a university. Begins to write papers and books—one thousand over his lifetime.
1805	Moves to Sicily where he spends ten years in business as a trader in natural products and secretary to the American Consul. Befriends the natural historian William Swainson with whom he explores Sicily. Publishes extensively, including the journal, *The Mirror of Science,* which existed for only one year, and *Fundamental Principles of Somiology,* dealing with classification of plants and animals.
1809	Marries Josephine Vaccaro. Daughter born in 1811; infant son dies in 1814.
1815	Publishes *Analyse de la Nature ou Tableau de l'Univers.* Unable to find a position in Italy. Returns to the United States. Shipwrecked off Long Island with the loss of almost all his collections and notes. Settles in New York City. Tours the Hudson River Valley and the Northeast. Helps found the Lyceum of Natural History of New York, the precursor of the New York Academy of Sciences.

1816 Publishes *Circular Address on Botany and Zoology.*

1817 Publishes *Florula Ludoviciana; or A Flora of the State of Louisiana.*

1818 Without prospects in New York. Leaves for Philadelphia and travels along the Ohio River to the Mississippi River. Visits Audubon and his friend and patron, John D. Clifford, in Lexington, Kentucky. Maps and describes Native American mounds.

1819 A busy winter in Philadelphia. Returns to Lexington as a Professor of Natural History at Transylvania University where he botanizes extensively, writes many papers, publishes the *Western Minerva,* and teaches natural history and several languages. Becomes interested in the history, languages, and archaeology of Native and Mexican Americans.

1820 Clifford dies. Rafinesque's position becomes increasingly difficult without protector. Several attempts to find new position meet with failure. Passed over by Jefferson's newly founded University of Virginia.
Publishes *Ichthyologia Ohiensis, or Natural History of the Fishes Inhabiting the Ohio River.*

1824 Engaged in historical and language studies of Native Americans. Attempts to establish a grand Public Garden in Lexington. Publishes *Ancient History or Annals of Kentucky.*

1825 Patents a *Divitial System* of banking and travels to major cities promoting his invention.

1826 Professorship at Transylvania University terminated; leaves Lexington and settles in Philadelphia, where he teaches and writes. Numerous travels on foot through the Northeast, collecting plants. In all names 6,700 kinds of plants.

1827 Announces the discovery and manufacture of *Pulmel,* which prevents and cures tuberculosis. Establishes a clinical practice as a *Pulmist,* carried out by mail.

1828 Publishes *Medical Flora of the United States.*

1829 Publishes *The Pulmist; or Introduction to the Art of Curing and Preventing the Consumption or Chronic Phthisis.*

1830 Publishes *American Manual of the Grape Vines and the Art of Making Wine*

1832 Publishes *Atlantic Journal, and Friend of Knowledge.*

1836 Publishes *The Life, Travels and Researches of C.S. Rafinesque.*

1836 Publishes *The American Nations; or Outlines of their General History, Ancient and Modern.*
Publishes *The World or Instability, A Poem.*
Publishes *New Flora of North America.*
Publishes *Flora Telluriana,* in 4 Parts.
Publishes the *Walam Olum,* the purported account of the wanderings of the Lenni Lenape or Delaware Indians from Asia to the shores of the Delaware River.

1837 Publishes *Safe Banking Including The Principles of Wealth.*

1838 Publishes *Genius and Spirit of the Hebrew Bible.*
Publishes *Ancient Monuments of North and South America.*
Publishes *Celestial Wonders and Philosophy, or the Structure of the Visible Heavens.*
Publishes *Alsographia Americana; or an American Grove of Trees and Shrubs.*

1839 Publishes *American Manual of the Mulberry Trees.*

1840 Publishes *The Good Book, Amenities of Nature or Annals of Historical and Natural Sciences.*
Dies of gastric carcinoma, September 18, in Philadelphia. His papers, books, and plant collections, many junked, sell for less than the cost of burial.

PREFACE

It was only a few years ago that the name Rafinesque meant nothing to me. However as I read early nineteenth century literature in biology and kept running across the name *Rafinesque* associated with the names of numerous plants and animals, I became aware that just pronouncing the euphonious name itself evoked the pleasant image of an exquisite French aristocrat. Darwin had acknowledged his contributions, and one could not explore great archives like those in the American Philosophical Society, the Academy of Natural Sciences of Philadelphia, or the University of Pennsylvania, without encountering *Rafinesque.* Similarly, he cropped up in numerous biographies of American natural historians of the time. As my reading in history, science, and biology broadened, I found that the name *Rafinesque* kept pace, for it was relevant in an astonishingly broad context, attested to by the fact that in the most complete bibliography of this man, there are over nine hundred papers of his on every conceivable subject. The more I read, the more I was drawn in; I had opened up a Pandora's box and out tumbled the encyclopedic cogitations of a mind on fire.

His discoveries and special insights were numerous and astonishingly wide-ranging, his writing, brilliant and provocative, all this emanating from a man whose behaviour was guaranteed to annoy, if not enrage, established authority. Bristling with confidence and incredibly well-informed, this natural scientist would not take criticism lightly, responding with curses and a renewed flood of information. Rafinesque's remarkable story has come out only in bits and pieces over the past one and a half centuries, and so I felt that a comprehensive presentation of the man and his work would help us appreciate him in the context of early nineteenth-century intellectual endeavors in the United States.

Just as previous biographical work depended heavily on Rafinesque's autobiography *A Life of Travels,* so does mine.[1] Those who have written about

Rafinesque's work have confined themselves to only one, or a few, of the fields in which he was active. Most have been scientists (botanists) without professional training as writers or historians. As a biochemist who has worked for over thirty years in the laboratory, I place myself in this earnest assembly. However, the present work gives an account of every area of Rafinesque's interests throughout his life with discussions of his papers and books, so that it is hoped the reader will acquire a more comprehensive view of his life, his work, and his time.

The short biographies of Rafinesque by R. E. Call and by T. J. Fitzpatrick have been extremely useful, as were the scholarly reviews and papers of Francis W. Pennell, Elmer D. Merrill, R. L. Stuckey, and Joseph Ewan. I am particularly indebted to the numerous, informative publications of Charles Boewe who has put the final touches on assembling a complete bibliography of Rafinesque, a dauntless task that was begun by Call and Fitzpatrick.[2]

I am grateful to the many librarians and archivists who have been so generous in their help: Donald Ewert; graphics artist Christine DeLaurentis; librarian of the Wistar Institute Nina Long; Carol Spawn; R. S. Ridgely and Robert Peck of the Academy of Natural Sciences of Philadelphia; Roy Goodman of the American Philosophical Society; Diane Cooter of the Syracuse University library; Catherine N. Norton, Jean Monahan, and Heidi Nelson of the Marine Biological Laboratory at Woods Hole, Mass.; Rita Docherty and Susan Glassman of the Wagner Free Institute of Science of Philadelphia; Gina Douglas of the library of the Linnean Society of London; University Archivist of Transylvania University B. J. Gooch; John Pollack of the Annenberg Rare Books and Manuscript Library of the University of Pennsylvania; Gretchen Worden of the College of Physicians of Philadelphia; and the librarians of the Gray Herbarium Library of Harvard University, the Historical Society of Pennsylvania, the Library Company, and the Bibliothèque Centrale du Muséum National d'Histoire Naturelle in Paris, France. Martin Cherkes of Princeton University was very helpful in banking matters and A. J. Stunkard of the Department of Psychiatry of the University of Pennsylvania School of Medicine provided much needed guidance.

I would like to thank Russel E. Kaufman and Clayton Buck of the Wistar Institute of Anatomy and Biology and Charles Emerson, chairman of the Department of Cell and Developmental Biology of the School of Medicine, University of Pennsylvania, for their unstinting support. As ever, I am especially grateful to my wife Eve for her careful editing of this manuscript and for always getting things right.

INTRODUCTION

In the eighteenth and early nineteenth centuries, just as creditable scientific studies were emerging in America, many extraordinarily creative people found their way to the United States from Europe. None was more perplexing and exceptional than Constantine Samuel Rafinesque, a force of nature who deservedly has had more written about him than almost any other nineteenth-century American natural scientist, and in some circles he has become an intriguing cult figure. A poet and a philosopher, he was the first professor of Natural History in the Midwest—the American frontier—and here he did his classic work on the fish of the Ohio River. Americans have been particularly beguiled by this adopted son; the very poetry of his name, his foreignness, and his singularity of deportment have commanded their attention. Even his severest critics have acknowledged his genius. Not unlike the poet William Blake, Rafinesque was "infected with the suspicion of insanity," and he often had "unleashing visions of another world."

A relentless scrivener, the sheer volume of Rafinesque's work, its variety and boldness in his one thousand publications, its grand ideas and insights freely expressed, his fantastic histories, left his peers dumbfounded and his biographer laboring to do justice to it all. His discoveries were numerous; his writing, often astounding and provocative. Rafinesque was intensely interested in the classification of plants, having brought with him from France an advanced conception of taxonomy that he preached to skeptical Americans. After years of close observation of plants and animals in the field, twenty-seven years before Darwin's *The Origin of Species,* he wrote brilliantly of the formation of new species through constantly arising minute changes in living forms in different environments that in time would become sufficiently notable to justify the creation of a new taxon with a new specific name. He walked from New England to Virginia, and on numerous occasions he crossed the Allegheny mountains with his eyes fixed to the ground, searching for new plants and animals, and in so doing he named legions of them—6,700 species—even more than the great Linnaeus himself. His botanical interest extended to the practical for the betterment of humankind—he wrote authoritatively on the economics of agriculture and farming practices, providing valuable information on the cultivation of tea bushes in the United States, the growing of grapes, and the making of wine.

A lexicographer, fluent in a remarkable array of languages, he published extensively on the languages of Native Americans and devised a statistical method of determining affinities between these many languages, thus classifying languages as he did plants. He wrote at length on the histories of Mexico and of Native Americans, and he was immersed in archaeology, examining and cataloging the mysterious mounds left by aboriginal people in America. Rafinesque was the first to attempt to interpret ancient Mayan hieroglyphics, and this he did with some success, discovering the nature of the numerical system used in their script, correctly concluding that the ancient Mayan language was related to its modern counterpart.

Rafinesque read exhaustively about all these matters in many languages, so his knowledge was unparalleled. The problem was that he sailed beyond factual knowledge and created his own history. Indeed, in all of his intellectual pursuits, the fine line between observation and imagination, between what he believed to be and what he saw, was sometimes conflicted—a shifting boundary that fascinates us. A man of confounding contradictions, he was a rationalist and a materialist, who at the same time exhibited mystical traits.

Profoundly humane, a bold defender of the scorned Native American, he considered slavery of any kind to be an abomination. Skeptical of the established medical practices of his day and especially scornful of heroic medicine—the bleeding of patients and the use of dangerous mercury potions—he chose an herbal approach to healing. Though not a licensed physician, he practiced medicine by mail, writing sensible, reasoned advice and prescribing herbal mixtures that at least did no harm. His *Medical Flora* or *Manual of the Medical Botany of the United States of America* was authoritative and became popular with both physicians and laymen.

Without question, botany was his passion, and in his *New Flora* he left a charming account of his explorations that reveals his ecstasy in the silent isolation of the American wilderness:

> Let the practical Botanist who wishes like myself to be a pioneer of science,

and to increase the knowledge of plants, be fully prepared to meet dangers of all sorts in the wild groves and mountains of America. The mere fatigue of a pedestrian journey is nothing compared to the gloom of solitary forests, when not a human being is met for many miles, and if met he may be mistrusted; when the food and collections must be carried in your pocket or knapsack from day to day; when the fare is not only scanty but sometimes worse; when you must live on corn bread or salt pork, be burned or steamed by a hot sun at noon, or drenched by rain, even with an umbrella in hand, as I always had.

Mosquitoes and flies often annoy you or suck your blood if you stop or leave a hurried step. Gnats dance before the eyes and often fall in unless you shut them; insects creep on you and into your ears. Ants crawl on you whenever you rest on the ground, wasps will assail you like furies if you touch their nests. But ticks the worst of all are unavoidable whenever you go among bushes, and stick to you in crowds, filling your skin with pimples and sores. Spiders, gallineps, horse-flies and other obnoxious insects will often beset you, or sorely hurt you. Hateful snakes are met, and if poisonous are very dangerous, some do not warn you off like the Rattle-snakes.

You meet rough or muddy roads to vex you, and blind paths to perplex you, rocks, mountains, and steep ascents. You may often lose your way, and must always have a compass with you as I had. You may be lamed in climbing rocks for plants or break your limbs by a fall. You must cross and wade through brooks, creeks, rivers, and swamps. In deep fords or in swift streams you may lose your footing and be drowned. You may be overtaken by a storm, the trees fall around you, the thunder roars and strikes before you. The winds may annoy you, the fire of heaven or of men set fire to the grass or forest, and you may be surrounded by it, unless you fly for your life.

You may travel over a[n] unhealthy region or in a sickly season, you may fall sick on the road and become helpless, unless you be very careful, abstemious and temperate.

Such are some of the dangers and troubles of a botanical excursion in the mountains and forests of North America. The sedentary botanists or those who travel in carriages or by steamboats, know little of them; those who merely herborize near a city or a town, do not appreciate the courage of those who brave such dangers to reap the botanical wealth of the land, nor sufficiently value the collection thus made.

Yet although I have felt all those miseries, I have escaped some to which others are liable. I have never felt compelled to sleep at night on the ground, but have always found shelter. I have never been actually starved, nor assailed by snakes or wild beasts, nor robbed, nor drowned, nor suddenly unwell. Temperance and the disuse of tobacco have partly availed me, and always kept me in health.

In fact I never was healthier and happier than when I encountered those dangers, while a sedentary life has often made me unhappy or unwell. I like the free range of the woods and glades, I hate the sight of fences like the Indians! The free constant exercise and pleasurable excitement is always conductive to health and pleasure. . . . What a delight to meet with a spring

> after a thirsty walk, or a bowl of cool milk out of the dairy! What sound sleep at night after a long day's walk, what soothing naps at noon under a shaded tree near a purling brook!
>
> Every step taken into the fields, groves, and hills, appears to afford new enjoyments. Landscapes and Plants jointly meet in your sight. Here is an old acquaintance seen again; there a novelty, a rare plant, perhaps a new one! Greets your view: you hasten to pluck it , examine it, admire, and put it in your book. Then you walk on thinking what it might be, or may be made by you hereafter. You feel an exultation, you are a conqueror, you have made a conquest over Nature, you are going to add a new object, or a page to science. This peaceful conquest has cost no tears, but fills your mind with a proud sensation of not being useless on earth, of having detected another link of the creative power of God. . . . When nothing new or rare appears, you commune with your mind and your God in lofty thoughts or dreams of happiness. Every pure Botanist is a good man, a happy man and a religious man! He lives with God in his wide temple not made by hands.[1]

There is a ring of authenticity in this field-worker's transcendent credo, a purity of heart and a deep honesty that anyone can appreciate.

Rafinesque, who came from a mercantile family, showed considerable entrepreneurial skills while in Sicily; and later in the United States, he formulated an innovative, transparent system of banking based on the humane principle that the small investor must be protected and should benefit as much from his investment as do the wealthy and powerful. Unfortunately, greedy bankers and entrepreneurs corrupted this novel system.

This list of extraordinary accomplishments is only half of his story. Alas, he was only a mortal who could not embrace the stars. His science was at times careless, and his writings were so assertive and on so many subjects that they could not help but be flawed. His mind, often fevered, outran normal capabilities, and ultimately his imaginative creations were adrift from the objective discipline of the real world. Such transgressions were eagerly pounced upon by his detractors. In virtually every field in which he was engaged, he provided a flood of information and speculation, much of it sound but some tragically flawed. He often went so far beyond the accepted views of his peers that he was denounced and ridiculed. Those who railed against him were especially angered by his severe and impolitic judgments of nearly everyone's work, including that of his friends and benefactors; he was his own worst enemy.

But as much as his judges fumed, Rafinesque would not concede a point to his supposed intellectual inferiors, and with the heat of argument rising, he would denounce those who had become his foes. He cursed his many tormentors—those who had cheated him in Sicily, those who did not give him proper credit for his discovery of new species of plants and animals, and the burghers of Lexington, Kentucky, who did not appreciate his efforts to increase knowledge for its own sake. Rafinesque was accursed in his private as well as his public life, and bad luck, like the hounds of heaven, seemed to pursue him relentlessly. There is enough tragedy in his story, both personal and professional, to provide Puccini with a dozen operas.

There always have been and always will be defenders and detractors of Rafinesque's reputation, and the proportion of each has varied as the years have passed. The detractors, many of whom had personal contact with him, predominated at the time of his death. These men had suffered not only Rafinesque's arrogance but were witness to the calumny that he brought upon himself. However, as the nineteenth century came to a close, several prominent biologists began to see the good in Rafinesque's work, and they concluded that he had been unjustly maligned. Recognition of his priority in naming many plants and animals has been forced upon the scientific world, which was obliged to accept his names if it was to abide by its own indispensable rules of nomenclature. His reputation has been reestablished to some extent, and he has now attained a degree of respectability.

Fully understanding this man may always be beyond our reach, and nearly everyone's view of him requires correction as one tries to sort through the tangle of his research. A Chaplinesque man with a keen intellect, his lonely striving, fierce indominability, absurd confidence, and naive optimism mark him as something magnificent. He was the quintessential field-man in an age of field-men who roamed America observing Nature's resplendent creations. A child of the Enlightenment, there still remained in him the peculiar imagination of the Renaissance alchemist, a troublesome Paracelsus. With his blazing energy, he strove to impose an ordered understanding on whatever befell his scrutiny, certain that he could see farther than others and do what they could not. Through his vast knowledge and exhaustive scrutiny, he created a wondrous edifice that compels examination bit by bit. Rafinesque was unique; we find ourselves solicitous and protective of him, forever apologizing and excusing this latter-day Don Quixote because we insist that his achievements are astonishing and that he be given a fair hearing.

Chapter 1

IN THE BEGINNING

Constantine Samuel Rafinesque took his first breath of air on October 22, 1783, in a Christian suburb of Constantinople (modern Istanbul), an ancient city that gave him his name, and after an extraordinary life, he died in Philadelphia on September 18, 1840.

Constantinople had produced no ordinary human being. A brilliant, ambitious man of the highest intelligence with a fantastic imagination, but the barest of diplomatic skills, he lived his entangled life with ferocious energy. His main love was botany, and his most pleasurable occupation was the search for new plants in the wilderness of America. To do so he traveled. Indeed, travel was a dominant theme of his life, his refuge and comfort. He covered the globe "on horse back, with mules and asses, in stages, coaches, carts, waggons, litters, sedan chairs, sledges, railroad cars, &c., and even on men's backs. . . . By water I have tried canoes, boats, felucas, tartans, sloops, schooners, brigs, ships, ships of war, rafts, barges, tow boats, canal boats, steam boats, keel boats, arks, skows, &c"—everything but "camels and balloons."[1] The list marks the intensity with which he followed his numerous interests, all with the same concentration, extending to taxonomy, zoology, geology, agriculture, history, philosophy, medicine, banking, archaeology, linguistics, and poetry. An eyewitness on the frontier who wrote endlessly on many subjects, he left a vast literature that reveals his thinking, often far ahead of its time, and often unacceptable to his contemporaries.

His French father, Francois Georges Anne Rafinesque (1750–1793), headed the Levant branch of the firm of Laflèche and Rafinesque of Marseille. His mother, Magdeleine Schmaltz (1767–1831), whose family came from Saxony, Germany, was born in Constantinople and reared in Greece.[2] In the year of his birth, amid troubled times, his father thought it best to bring the family to Marseille, whereupon he returned to his business in the Middle East. In

Constantine's unusual way of seeing the world and of drawing odd conclusions from experience or supposition, he wrote: "I was yet in my cradle and at the breast of my mother, when my parents went with me to France by sea, by Smyrna and Malta where we stopt [sic] in the way to Marseille. The first and early voyage of mine, made me insensible or not liable ever after to the distressing sea sickness. By my observations ever since, it appears that whoever travels by sea in the cradle or very early, is never liable afterwards to this singular disorder. It seems also that whoever rides backwards in a coach without difficulty, is not liable to it; but whoever cannot, will suffer from it." Rafinesque must have believed this absurdity for a half century without questioning its veracity.

Constantine's early years were spent in Marseille with his mother and with his father's family. He was fluent in French and could probably speak his mother's Greek tongue. His earliest memories were of the beautiful suburbs of Marseille. "It was there among the flowers and fruits that I began to enjoy life, and I became a Botanist." As a boy, he won a book about animals, and so he wrote that the course of his life was set—traveler, botanist, zoologist, and naturalist. When he was six years of age, he and his parents traveled to Leghorn (Livorno), Italy, to visit his father's sister, who was married to an English merchant, Mr. Demaretz. The Mediterranean voyages remained vivid in the boy's mind, marking the beginning of his "personal observations in travelling."

A merchant of some substance himself, M. François Rafinesque, part owner of the trading ship, *Argonaute,* hoping to enlarge his fortune, set sail for Mauritius and China by way of the Cape of Good Hope in 1791. On the return voyage, he barely escaped from marauding British ships by seeking haven in the port of Philadelphia in 1793, which at the time happened to be paralyzed with a grave epidemic of Yellow Fever. Government officials, including Washington and Jefferson, had wisely fled the capital city, but the unfortunate M. Rafinesque sailed into this contagion. Scarcely had he sold his goods and the ship itself than he contracted the fearful disease and died along with five thousand other victims, which represented approximately 10 percent of the city's population.

In the early 1790s the horrors of the French Revolution were reaching new levels of intensity. The house of M. Laflèche, M. Rafinesque's partner, was burned by a mob, and by 1792 a fearful Madame Rafinesque, her husband halfway around the globe, thought it prudent to flee with her two sons, Constantine and the younger Antoine Simon Auguste (1785–1826), and her daughter Georgette Louise (1791–1834). They spent the next four years, from

1791 to 1834, in Leghorn with her husband's wealthy parents who had probably fled Marseille earlier. Constantine was taught by private tutors. He studied geography, geometry, history, drawing, and English, and he also became fluent in Italian. But from early on he went his own way, devouring books, especially those on natural history and travel, and he claimed that by the age of twelve he had read the "great Universal history, and 1000 volumes of books on many pleasing or interesting subjects"—perhaps a slight exaggeration. Trips to Pisa and Genoa and along the Arno River further inflamed his passion for travel, and his first substantive essay, at the age of twelve, was *Notes on the Appenines,* an account of a trip by mule and sedan chair from Leghorn to Genoa. In his reading, the voyages of Captain Cook, Le Vaillant, and Pallas were particularly exciting. He began to botanize in earnest, collecting plants, noting their habitats, and learning Latin so that he could read the publications of the previous century.

In 1797 he returned to his grandmother in Marseille, where he spent the next three years "to complete my education by myself." He read books of every description but seemed to specialize in those devoted to the natural sciences, medicine, moral philosophy, and chemistry. Constantine recorded that at that time, with reading and thought, a critical faculty was developed that enabled him to distinguish "good" buøks from "bad." He devoured a learned work such as Valmont de Bomare's six volume *Dictionnaire Raisonné Universel D'Histoire Naturelle,* each volume seven hundred pages long and crammed with information that began with *Aavora* and ended with a definition of *zygene.* The penultimate word was *Zurnapa.* (*C'est la giraffe.*)

He also buried himself in *Spectacle de la Nature,* written in the form of philosophic conversations between the Chevalier de Breuil (a "young gentleman of quality"), and the Count of Picardy as they walked about in the Count's garden. All of the supposed discourses were recorded by the Abbé le Pluche, a learned and pious gentleman, who presented the conversations as dialogues in a play. There were fifteen dialogues in all, and all were on natural history. Rafinesque relished the notion of being a chevalier and a nobleman's devoted student.

In his early teens Rafinesque was profoundly moved by the popular novel *Paul and Virginie,* written in 1788 by Bernardin de Saint-Pierre, whose characters were unblemished children of nature, living happily with their unwed French mothers on the exotic island of Mauritius, a botanical paradise in the Indian Ocean. They were destined to marry, but Virginie, forced to return to the civilized (and corrupt) society of France in order to inherit the family fortune, could not bear separation from Paul and the pastoral life she knew, and while returning to her Eden she perished in a shipwreck clutching a small portrait of her love. Not long after, Paul and the two mothers died of melancholy and despair, their paradise destroyed. This "simple," compelling melodrama was largely taken up with descriptions of flora, scenes of nature, and practical matters of plant cultivation. All were lovingly described in a world that was essentially run by women—idealized and yet revered in a condescending manner by the author. The happy group practiced a benign theology close to nature, only nominally Christian. The lesson to be learned was from Rousseau. Nature, which was all powerful—providing insight and happiness—was surrounded by human turmoil and the evil forces of civilization. Rafinesque believed that the scientist in nature was in a state of grace—righteous, happy, and basking in the glow of a benign, natural religion. The moral most certainly influenced and reinforced Rafinesque's nascent Romanticism, and as his own life unfolded he must have realized that it touched upon Paul's and Virginie's experience in many ways—his exaltation, alone in the great temple of Nature; his life outside France as an exile; his fruitless search for an ideal woman; and the shipwreck on an American coast that almost destroyed him, as it did Virginie on the shores of her Eden.

Although he valued formal education (for others) and taught in a university, he boasted: "I never was in a regular College, nor lost my time on dead languages; but I spent it in learning alone and by mere reading ten times more than is taught in Schools. I have undertaken to learn from the Latin and Greek, as well as the Hebrew, Sanscrit, Chinese and fifty other languages, as I felt the need or inclination to study them."[3] It would seem that no institution of learning could contain this remarkable mind and personality; perhaps they would not have him. The impatience, not to say arrogance, of this intelligent, overbearing boy is evident. An isolated, lonely soul, without friends, ignorant of games and play, considered strange—from such a boyhood emerged the man who never faltered in his belief that he could understand and master all that was known and unknown. He set his cogitations down on paper for the benefit of the less endowed, and at the end of his autobiography he wrote lofty words that bespoke a blinding confidence in his ability and capacity to achieve:

> Versatility of talents and of progressions, is not uncommon in America; but those which I have exhibited in these few pages, may appear to exceed belief: and yet it is a positive fact that in knowledge I have been a Botanist, Naturalist, Geologist, Geographer, Historian, Poet, Philosopher, Philologist, Economist, Philanthropist. . . . By profession a Traveller, Merchant, Manufacturer,

> Collector, Improver, Professor, Teacher, Surveyor, Draftsman, Architect, Engineer, Pulmist [lung specialist], Author, Editor, Bookseller, Librarian, Secretary. . . . [A]nd I hardly know myself what I may not become as yet: since whenever I apply myself to any thing, *which I like,* I never fail to succeed if depending on me alone, unless impeded and prevented by lack of means, or the hostility of the foes of mankind.[4]

Rafinesque's story provides ample evidence that he was enormously capable and excessively energetic—manic—with an unrestrained and undisciplined imagination. These characteristics he had as a boy and as a man. Unfortunately, in his formative years there were no restraining, calming, and guiding influences, so that by the time he reached adulthood he was a fearsome autodidact who lacked a critical sense of the limits of human capability—he was a loose cannon. If an element of paranoia surfaced, it was sometimes for good reason, for he made many enemies, some of whom were influential.

His lifelong passion for travel was proclaimed on the frontispiece of his autobiography *Life of Travels:*

> Un voyageur des le berceau,
> Je le serais jusqu' au tombeau . . .

Though he claims he could have entered any profession, he chose to be a merchant like his father because, as he said, "commerce and travel are linked," and he could still indulge his interest in botany. He began as an apprentice clerk for a distant relative, and in his spare time he expanded his studies to include fish and birds, shells, and crabs. He drew maps, copied rare works, made topographical surveys and wrote about geography. It was during this period that Constantine, in his midteens, began corresponding with the French zoologist, François Marie Daudin (1774–1804), sending him specimens and discussing taxonomy. Daudin was the first in a lifetime's harvest of professional correspondents. When very young, he expressed a desire to write his autobiography, but this did not occur for forty years, though not for want of experiences, opinions, and commentary throughout those years.

By the end of the French Revolution, Constantine's father and uncle were dead, and the remainder of the family fortune, entrusted to M. Laflèche, was dissipated and lost. Constantine was convinced that he and his brother were cheated, but the loss may have been due to hard times, great social upheaval, and a depression. Unfortunately, young Rafinesque no longer had an inheritance to draw upon and was thrust into the ranks of those who had to make their own way in the world. A dangerous journey took him from Marseille to his mother's home in Leghorn where he remained for two years working in commerce for his new stepfather, Pierre Lanthois—but he had time to study botany and the sciences. "I began to hunt, but the first bird I shot was a poor *Parus* [chickadee] whose death appeared a cruelty to me, and I have never been able to become an unfeeling hunter,"[5] a sentiment not shared by Audubon and many other naturalists, who had no compunction about shooting animals. Life was leisurely and pleasant visiting the surrounding countryside, gardens and museums, and the mountains of Tuscany.

In 1802 life began in earnest when it was decided that Constantine, who had always fancied himself the traveler in the style of Alexander von Humboldt, should seek his fortune in America. Prudence dictated that with the roiling instability of Europe and its constant wars, young men of military age should seek a pacific land. Furnished with letters of introduction, he and his younger brother Antoine set sail from Leghorn on the American ship *Philadelphia* commanded by a Captain Razer. Ever the observer, Constantine studied and drew fish, turtles, and molluscs he caught on the uneventful voyage. After forty days America came into view—the green, tree-lined shores of Capes May and Henlopen at the mouth of the Delaware River—and in another two days, in April 1802, the ship landed in the lively port of Philadelphia.

Young Constantine and his brother Antoine (Anthony Augustus), old-world sophisticates, had journeyed to a new kind of country, a pastoral Republic whose president was Thomas Jefferson, an intellectual agrarian with reverence for the rural life and who was suspicious of both centralized authority and the manufacturing and merchant classes, with their Federalist tendencies. He envisioned an America of "honest farmers and country gentlemen," dependent upon the export of farm products and the import of manufactured goods. The expansion of urban centers, with their factories, ignorant masses, corruption and disease, disturbed this president. Though Rafinesque settled in the city, he roamed the land, where he encountered Jefferson's country gentlemen everywhere, for at the time nine out of ten Americans lived on farms.[6]

The America to which Rafinesque came was one of untold natural riches, an almost unexplored, unspoiled paradise that was unimaginably vast and laden with promise and at the very beginning of its spectacular awakening in industry and manufacturing. Building a new country that would be the envy of tired, old, monarchical Europe was America's task. Citizens were concerned with commerce, the growth of cities, the settlement of land, and uppermost, the making of personal fortunes. Since the country's wealth was modest, capital needed to build roads, railways, and canals came flooding in from Britain and the rest of Europe. Though it had a few institutions of learning, museums, and libraries, they were notably inferior to those in Europe, and even in 1848 an American scientist visiting London could write: "All our museums sink into entire insignificance when compared to those existing here."[7] American distinction lay in its fathomless natural resources of unspoiled, wondrous variety, its primitive native inhabitants, and its abundance of unknown plants, animals, and minerals.[8] John Locke, seeing Europe's distant past in America, was prompted to say, "In the beginning all the world was America," and Europeans were enchanted by firsthand reports of the New World, a paradise.[9]

Chapter 2

THE NEWCOMER IN BOTANICAL PARADISE 1802–1805

> When he walks forth to enjoy the beauties of nature, not a butterfly can wanton in the sunbeams; —not a flowret can raise its modest head; —not a flash of lightning can coruscate in the Heavens; —not a fish can leap above the glassy bosom of the streamlet; —not a particle of 'atmospheric dust' can fall upon his coat, without filling his breast with sensations of delight: *all are caught, examined, classified and described.*[1]

Portentously, as the nineteen-year-old Rafinesque stepped off the ship onto American soil he recorded: "The first plant that I picked up was also a new plant, then called *Draba verna,* and that I called *Dr. Americana,* altho' the American Botanists would not believe me; but Decandole [de Candolle] has ever since made with it the new Genus *Erophila*! this is the emblem of many discoveries of mine, of which ignorance has doubted, till science has prove that I was right."[2] The "new plant" he had reclassified was, in fact, *Draba verna,* a cruciferous plant, a common, well-known weed in Europe and America. He believed he saw morphological characteristics of the plant that distinguished it from the European *Draba verna,* making it a new species that justified a new name, but detailed researches now indicate that he was in error. Rafinesque held the general view that any plant found in America could not be of the same species as its European counterpart. The incident, however, confirmed the young man's belief that American botanists, living on a half-known continent abounding in undescribed plants and animals, were deficient in their ability to classify them properly and that his mission was to edify the natives.

Philadelphia at the time of Rafinesque's arrival was a market center and seaport of more than 70,000 citizens that saw some 3,500 ships arrive and depart yearly. Trade and commerce thrived here—the largest city in America until overtaken by New York in 1810—although the population of the entire country was mostly rural and agrarian. Philadelphia was the publishing center of the country and could boast of the richest cultural life, graced with theater, music, the American Philosophical Society, and Charles Willson Peale's Museum of Natural History.[3] The latter, a "closet of natural curiosities," proved to be an important institution of learning for many fledgling natural scientists of Philadelphia, including Rafinesque. In his earliest American papers, Rafinesque described specimens of four species of birds from Java, found in the museum.[4] The year of his arrival was marked by the sensational exhibit of a reconstructed, fossil mastodon excavated by Peale in Newburgh, New York, in 1801. The exhibit, a godsend for Peale's finances, was visited by multitudes, including the most illustrious—Washington, Jefferson, Alexander Hamilton, Robert Morris, and James Madison. The city could boast of the University of Pennsylvania with the finest medical school in the land, where, despite his distrust of large cities, Thomas Jefferson planned to send his grandson for instruction.[5]

Rafinesque had come to the right American city, for Philadelphia was a cosmopolitan community, able to proffer a measure of European refinement. Men and women of intellectual distinction—émigrés, professionals, naturalists, scientists, and merchants—sympathetic with Enlightenment and Utilitarian principles, were able to provided the stimulating companionship and conversation that Rafinesque craved. First and last, his major passion was botany—"my favorite science, the most amiable of all"—and his newly adopted city had been the center of botanical studies in the country for more than fifty years. If there were any city that abounded in the requisite talent and facilities to assemble a *flora* of American plants it was the Quaker city, which was so rich in gardens, plant collections, and botanical libraries. The mid-Atlantic area had given rise to a broad community of scholars and natural historians passionately interested in plants, who collected, cultivated, and classified them, and were eager to share their knowledge.[6]

The Quaker, John Bartram, had established the first important botanical garden in America in 1730 and had extensive connections with European botanists and horticulturists whom he supplied with plants and seeds for medicinal, agricultural, and decorative purposes. He, and his son William explored the eastern states from New York to Florida, collecting seeds and specimens, many for shipment to England. In their garden grew *Franklinia alatamaha,* a lovely tree they had discovered confined to a river valley in Georgia. The tree, named after Franklin, could not be found on a later trip and is now considered extinct in the wild.[7] A second important botanical garden was established in

1773 by John's cousin, Humphry Marshall, who wrote the first botanical work ever published in America, a complete account of native trees and shrubs. A third garden, *The Woodlands,* near Bartram's garden and adjacent to the present site of the University of Pennsylvania, was established by the wealthy William Hamilton and was unlike anything to be found in America in its magnificence and comprehensiveness. Through Hamilton's efforts, the gingko tree and the Lombardy poplar were introduced into the United States.

The city of brotherly love proved welcoming to the Rafinesque brothers. Typical of the time, the kindliness shown to Europeans was documented in de Crèvecoeur's *Letters from an American Farmer.*[8] The Cliffords, a family of merchants who owned the ship on which the Rafinesques had arrived, hired them as clerks in their counting house. John D. Clifford, whom Rafinesque had met in Italy in 1802, was also an inveterate wanderer with a strong interest in natural history and Indian relics. Clifford became a good friend and patron of Rafinesque, and it was he who probably induced Rafinesque to come to Philadelphia. Benjamin Rush, signer of the Declaration of Independence, considered by some to be the most famous physician in the United States, was particularly solicitous, offering Constantine an apprenticeship in his practice. Constantine preferred commerce to Medicine, not only because he was genuinely interested in business with its attendant travel, but also because the demands on his time would not be as great—time that he could spend botanizing and exploring.

Summer, with its heat and the dreaded Yellow Fever, was soon upon the city, and Rafinesque, mindful of his father's misfortune in Philadelphia, fled to the safety of Germantown (now part of the city). Here he befriended a Colonel Thomas Forrest, a veteran of the Revolutionary War and an eccentric "friend of Horticulture," who had begun his career as an apothecary with some knowledge of medicinal herbs and plants. He ended his career as a congressman from 1819 to 1823 walking about Washington in full Quaker garb, switching back and forth between political parties. Rafinesque lived and traveled with the colonel, collecting and naming plants and describing birds, reptiles, and fish.

A pattern of tireless exploration and collecting began. They examined the area around Germantown and the Pine Barrens of New Jersey and visited the Marshall Botanical Garden in West Chester (Pa.) and John Bartram's famous garden on the banks of the Schuylkill River. Perhaps these gardens and conversations with local botanists began to change his mind about American capabilities. Upon seeing his "first Indians or ancient natives" his mind was fired up as he pondered their origins and he began to study their language, cultures, and history.[9]

In October when the epidemic had receded, Rafinesque settled down to a winter of commerce in the city, but the counting house rapidly lost its appeal, because the demands on his time were far greater than he had anticipated. As a young man of twenty, he was too restless, too much the wanderer to be confined to a sedentary clerkship. He had visions of being a sea trader like his father, but "I was yet too young to be entrusted with a voyage."

The year 1803 saw the return of Yellow Fever, and again he fled to Germantown, this time accompanied by his brother Antoine Auguste, who had spent his time unemployed in New York and Newark. Constantine seemed to have little brotherly concern, for he rarely talked about Antoine Auguste in his autobiographical accounts. Indeed, he seemed to have had little active regard for any member of his family. He rarely mentioned his father, mother, and only sister, and it is almost incidental that we have learned that he was married in Sicily and had fathered two children.

Once again Rafinesque began his botanical journeys, "pedestrian excursions," throughout Pennsylvania and New Jersey, his forays taking him farther and farther into the wilderness. Insisting that he be near the ground so that he could examine plants more closely, he chose to walk rather than travel by horseback, for frequent dismounting was tiring, and so he wrote "horses do not suit botanists."

Rafinesque spent some time in the iron mines in Cornwall, Pennsylvania, the Moravian college in Ephrata, and the Blue Mountains. He examined the flora and fossils found along the Juniata River north of Harrisburg. In the course of his wanderings he visited many botanists around Philadelphia, and not confining his interest to plants, he also collected and studied snakes and reptiles, reporting his observations to his old correspondent, the zoologist François Daudin, in France. He was in fact, one of America's first herpetologists. Finding many unfamiliar and "unknown" birds, or birds that were "badly described," he planned to, but never did write an ornithology of the United States.

By 1804, impatient with his sedentary job at the Clifford establishment, he had decided to give it up and devote all of the next year to botany and the exploration of America, and then he would return to his homeland:

> My pedestrian excursions of the last year had given me a relish for these rambles; I had become convinced that they were both easy, useful and full of pleasure, while they afforded me the means to study everything at leasure[sic]. I never was happier than when alone in the woods with the blossoms. or resting near a limpid stream or spring. I enjoyed without control the gifts of

> Flora, and the beauties of Nature. I therefore resolved to undertake this year, longer journeys before I left America, where I foresaw that I could not remain to advantage, as I often threw my eyes towards Greece and Asia, as another field of exertions and discoveries.[10]

He began to read accounts of travels and explorations in America, and the history of the continent. His brother filled the vacancy at the Clifford counting house while he headed for the open road, the forests and fields of the mid-Atlantic states, bearing letters of introduction to the most prominent citizens of every locality, which assured him a cordial reception. Rafinesque explored the Atlantic shores of Delaware and Maryland and the area around the Delaware Water Gap, feverishly collecting specimens and making himself known to important citizens. What he lacked in humor, he made up for in charm, and however odd he may have seemed, he was a pedant worth listening to. The list of his contacts reads like the Who's Who of American botanists. All in all, he walked 1,200 miles on this field trip.[11]

In the early summer of 1804, he made the requisite pilgrimage to Washington, and by chance, he came upon a deputation of Osage Native Americans who were visiting the capital. Rafinesque was presented with an opportunity to study a reasonably intact culture of aboriginals at close hand, earnestly observing their ceremonial dances, and learning something of their language through an interpreter.

Rafinesque seemed to know everyone. He had many important friends and contacts, and armed with a letter of introduction from the senator from Pennsylvania, he met Secretary of State James Madison and President Jefferson, and through the secretary of war he met General Henry Dearborn. Jefferson, a Francophile, was impressed by the brilliance and erudition of this young European, despite Rafinesque's bad habit of interrupting and even correcting the president. Both men were widely read, with broad interests, especially in the natural world, but unlike Rafinesque, Jefferson had a disciplined mind and was exquisitely political. Rafinesque's intimate and prodigious knowledge of botany, both theoretical and applied, could be of use to the president, especially because of his great concern with practical gardening and agriculture, particularly as practiced in France and Italy. A few days after their meeting, Rafinesque visited with a gift of seeds of *Jeffersonia diphylla*, only to find that the president had left for Monticello. The two men met only once.

A few cordial letters were sent over the next six months.[12] Rafinesque wrote about exploring as far as "the Blue-mountains of Pennsylvania and New Jersey," and about his plan to travel through Virginia, Kentucky, and Ohio. He regretted that the western parts of the country (of special interest to the president) were virtually unknown, especially across the Mississippi River. If the government would organize an expedition to these parts "I would think myself highly honored with the choice of being selected to make known the Vegete and Animal riches of such a New Country and would think that Glory fully adequate to compensate the dangers and difficulties to encounter."[13] Jefferson replied within a month that money was being requested from the legislature to finance a mission to the Red and Arkansas Rivers, and if successful, Rafinesque could serve as botanist. However, members of the expedition would not be paid.[14] Unfortunately, a despondent Rafinesque had set sail for Leghorn, Italy, a few days previously, so once again, his timing was off. Perhaps his life would have been radically different if he had been able to accept the president's offer,[15] but it is also possible that the expedition would have proven a disaster for Rafinesque because he was not a team man. Even before the Red River–Arkansas mission, Rafinesque had hoped to accompany Lewis and Clark on their expedition to Oregon as a botanist and surveyor, but another highly qualified man, the ornithologist Alexander Wilson had also been passed over, so he didn't even apply. Perhaps in an enterprise such as this, President Jefferson preferred hardy, native-raised American hunters and explorers of a practical outlook rather than specialized, theoretical Europeans.

On his visit to Washington, he had explored the surrounding area and had visited numerous towns and cities—Alexandria, Baltimore, Harrisburg, Reading, and Bethlehem among others—assiduously collecting specimens and garnering information about local flora and fauna from knowledgeable citizens. He summarized his botanical activities in and around Washington and Delaware in manuscripts entitled *Florula Columbica* and *Florula Delawarica,* which were catalogues of plants he had collected. Neither of these works was published, though Rafinesque frequently mentioned them and listed them as his first botanical studies in America.[16] The manuscripts had been sent to Dr. Benjamin Smith Barton, Professor of Materia Medica, Natural History, and Botany at the University of Pennsylvania, and editor of *The Philadelphia Medical and Physical Journal,* where it was announced in print that their publication would be forthcoming—but it never appeared nor was the manuscript returned.[17] In his bibliography, Rafinesque frequently stated that the work, listed as his fourth, was "suppressed" by Barton, and this would appear to be true. Barton may have been attempting to commandeer Rafinesque's work for his own large botanical opus.[18]

This egregious affair may reflect the growing distrust and open hostility that the local botanical community had for Rafinesque who was regarded as a hyperkinetic upstart, and it must have suggested to Rafinesque that publication in American journals would be increasingly difficult, if not impossible. The very large number of new plant species that Rafinesque hastily described (often without acknowledging the sources of his botanical material) astounded his fellow botanists. He had the irritating habit of announcing the imminent publication of works so overly ambitious that they could not possibly be brought to fruition. G.H.E. Muhlenberg, wrote to a friend: "Have you seen what Mr. Rafinesque Schmalz has printed in the New York Medical Repository and what he promises to publish hereafter? He makes a wonderful change and havoc amongst our plants and will do much harm if he keeps his promise. I know him personally and find a great number of my plants, which I gave him, superficially described without mentioning a word from whence he had them. Very often he makes a genus where hardly a species can be made, and where his specimen was quite imperfect. There is a medium in everything. In botany the *festina lente* is very necessary."[19]

Rafinesque's inspiration was Linnaeus, whose dictum was that it was the duty of man "to affix to every object its proper name." His work was born of genius, and Rafinesque would attempt to emulate and perhaps match his accomplishments. Rafinesque's mentors were all eighteenth and early nineteenth-century Europeans, mostly French—de Jussieu, Ventenant, Adanson and Necker in Botany, and Lamarck and Daudin in Zoology, and there were many encyclopedias and published works, in both English and French, from which Rafinesque borrowed. He was a born classifier and systematizer of knowledge, immersed in the Enlightenment tradition of the great savants—*philosophes*—who occupied themselves compiling dictionaries, encyclopedias, and handbooks in their attempt to inventory all knowledge and give it structure—a monumental task to which Rafinesque eagerly dedicated himself. In accordance with the principles and spirit of the Enlightenment, he believed that human judgments must be based on observation and reason, and that ultimately, in an ideal world, an integration of science, art, and philosophy would be achieved. Rafinesque would impose an order, a system of classification, on any set of facts derived from nature or society as attested to by his *Analysis of Nature, or Tableau of the Universe and of Organized Bodies* or his *Fundamental Principles of Somiology.* Yet despite all his rationality, it is not difficult to find a mystical, Romantic strain in the writings of Rafinesque, who was the child of an age in which art, literature, natural philosophy, and social theory were regarded as elements of an integral whole. But beyond this, there is a characteristic peculiar to Rafinesque that distinguished him from his peers—the blind acceptance of information, often false and absurd, which in the end corrupted so much of his opus. Rafinesque stood by his immutable internal verification system and framework of beliefs that ran afoul of the dictates of rational analysis and objective verification.

The habit of preserving specimens and building collections, selling or trading extra specimens, was practiced from the late eighteenth century on, with private collections often finding their way into permanent homes by donation to museums and institutions, the first such American institutional herbarium being that of the Academy of Natural Sciences of Philadelphia (1812).[20] Not only did preserved specimens please the eye, they became type specimens, standards to which other plants could be compared, identified, and named. A serious deficiency in early American botany was the lack of such botanical standards, and so American botanists were dependent on European herbaria. The need for American collections was soon recognized and in part remedied by such men as John Torrey and Asa Gray, spurred on by a growing nationalistic feeling.

Since the New World abounded in plants that were unknown to Europeans, American botany soon took on an international character. Amateurs and professionals on both sides of the Atlantic wove a network, and in time, with the constant accretion of information, rules of behavior and an historical perspective evolved to impose an order on the enterprise. Botanists were constantly collecting and distributing specimens to their colleagues in all parts of the world, and all willingly subscribed to commonly accepted rules such as priority in the naming of species and genera (determined after 1867 by the International Code of Botanical Nomenclature). Lack of adherence to accepted rules would lead to confusion and chaos, for in taxonomy, which Rafinesque called "the branch which teaches us the alphabet of the science," the record of the past is continuous with the operations of the present, and error accumulates and becomes magnified, creating an unstable house of cards. Some botanists were more difficult to keep in check than others, and among the special cases, Constantine S. Rafinesque was notable. Though generous in his sharing of information and specimens, he described numerous pseudospecies and genera, and his identifications were not definitive. This headstrong "species splitter," a creator of new species on the slightest of differences, real or imagined, disregarded the rules of classification and consciously or not, mocked the orderly and widely recognized system.

Various imperfect schemes for the classification of plants and animals had been devised and pondered over for centuries. A revolutionary advance was made in 1736 with the publication of Linnaeus's *Systema Naturae* in which rules for classification were established. The scheme was a triumph of rationalist thinking in which three realms of nature—plant, animal, and mineral—were outlined.[21] The work was, in essence, a practical, concise guide for identification of living forms and for their placement into categories. In the plant kingdom, classification was based on the reproductive parts—the number and position of the stamens and pistils of the flower—and from this information, in sequential steps, the higher categories of classification of members of the plant kingdom could be determined, with class from the stamens and order from the pistils. A binomial system was devised in which every organism was given two names, a *generic* and a *specific* name. Species bore characteristics that were unique to an organism, meriting a *specific* name. Different species possessing certain broad characteristics in common were classified as members of a genus.[22] In effect, a hierarchical sequence was constructed that was called the *artificial or sexual* system of Linnaeus.

Modern botanical nomenclature dates from 1753, with the publication of Linnaeus's *Species Plantarum* in which 5,800 plants were listed, each with a binomen—a generic and a specific name.[23] To make identification convenient, catalogues of binomens were compiled of each name, followed by a short description of the distinguishing features and habitat of the plant or animal in question. Definitions and rules of nomenclature, appropriately amended, have remained universally employed to this day. Fortunately, they were adopted and discipline was imposed at a time of impending chaos when naturalists in every country and those on global expeditions were discovering large numbers of new plants and animals that would require definitions and names. Since much of the confusion and complexity of identification was eliminated by an orderly arranging of taxa, botanizing and the collection and preservation of plants became an avocation that has been enjoyed by multitudes of naturalists.

The *artificial* or *sexual* system of Linnaeus had the virtue of simplicity, and though faulty in many ways was useful and widely accepted, especially by the Germans, Scandinavians, and British, and through them by Americans. However, French naturalists questioned why only the sexual components of a plant and not other parts should be the basis of classification. Though reproduction is a fundamental, critically important process, it seemed arbitrary to assign an overriding priority to the sexual apparatus. Linnaeus himself recognized the imperfection and the complexity of the problems involved, stating that anyone who could resolve the problems of classification would be his *great*

Apollo, and indeed, there were alternative ways of dealing with the question of nomenclature and the organization of taxonomic groups that were based on a broader range of criteria.

Toward the end of the eighteenth century, the *artificial* system of classification and the supposed primacy of sexuality were challenged by the *natural* system, a French creation whose leading proponent was Antoine-Laurent de Jussieu, a member of a family of botanists.[24] In this method of classification, a greatly increased number of discernable characteristics of plants (not just the sexual) were compared and correlated. Taxonomy would depend upon as many characters as could be defined, although de Jussieu did establish a hierarchy, some characteristics being more important than others. Sexual characteristics were indeed most important for classification, followed by those that were critical for the plant's survival or those that were most frequently found, so that a whole range of major characteristics of a species or genus was essential for assigning members to a taxon. Later, function and biological relationships were incorporated into the process that made classifying more precise but more difficult and complex. Presciently, Rafinesque criticized the work of Thomas Say and others for neglecting physiological characteristics in their descriptions,[25] and yet paradoxically, though he preached the *natural* system, Rafinesque was critical of those who went so far as to dissect fruits and seeds and finely analyze the structure of flowers. He believed that the *external* characteristics of the flower, fruit, and seed were sufficient *in all instances* to achieve a complete classification, and there was no more reason to look *into* a seed than there was to look *inside* an egg to determine the kind of bird it would give rise to. Working with too many characteristics would only lead to confusion, more than the mind could handle. Being in a hurry, in practice Rafinesque found himself "guided by gross similarities" both for plants and animals.[26]

Still, on the expanded, *natural* basis, new arrangements of families, genera, and species were devised, and a sense of relationships and relative importance of various characteristics emerged so that the number of "correct" placements was greatly increased, but the underlying struggle between the dictates of reason and of intuition was always in play, and remains an important factor to this day. Systematics at that time had been based on similarities of characters without a theoretical foundation for their use. Plants with, say, white flowers or with a particular shape of leaf were grouped together. Some believed that no relationship or kinship existed between living forms, for each was created separately and independently and remained immutable in accordance with the inscrutable plan of the Creator, and so species could simply be listed alphabetically.

A rational basis for discerning priorities and relationships would be provided by Charles Darwin's *On the Origin of Species by Means of Natural Selection* (1859), which posited the descent of living forms from a common ancestor through an evolutionary process driven by natural selection.[27] Classification of plants by the natural system came to be dependent on ancestry, and the phylogenetic tree that outlined the relationships among the plants (and animals) defined by taxonomy would be nothing less than the roadmap of evolution. Although the complete description of a living form was "intricate," the intricacy had to be accepted, for nature was indeed just that complex.

In the process of classification, characteristics, mainly anatomical and seen with the naked eye, predominated. But with the introduction of the microscope, a whole new range of attributes, especially needed with small organisms and invertebrates, could be exploited. The wealthy businessman-geologist William Maclure had provided the Academy of Natural Sciences of Philadelphia with a compound microscope in 1817, but such were the times that members of the Academy, including Rafinesque, rarely took advantage of this powerful instrument, for they thought it would only provide useless detail that would be more confusing than edifying. Despite Lindley's successful promotion of the Natural System of classification in Britain, Rafinesque disliked his seminal book on the subject, because Lindley espoused the use of microscopic characteristics in the classifying process, among other objections. Though Rafinesque craved discovery and new knowledge and his curiosity was remarkable, oddly enough he was often relatively narrow in his approach to the study of man and nature. He observed, described, and classified, and he brought imagination and vast erudition and experience to his studies. But he did not take up the microscope or any other instrument that would have intruded between his eyes and the object of study. Experimentation would have also expanded Rafinesque's understanding of nature, but he was never interested in this manipulative process and the times did not demand a new approach to discovery. As it was, Rafinesque had more than enough to occupy his days and nights.

Many leading botanists, especially in France and Switzerland, rapidly accepted the natural system. Correa da Serra, the Portuguese consul in Philadelphia and a botanist, had lectured on the new system in 1815,[28] as did the young Rafinesque, but for decades their discourses fell on deaf ears, especially those of the older members of the Academy of Natural Sciences of Philadelphia. American botanists were dedicated to the *artificial* or *sexual* Linnaean method of classification—much to their detriment. Thomas Jefferson had an understanding of the principles of classification and of the *sexual* system, but did not seem to be concerned with, nor was he even aware of the conflict between this system and the newer, *natural* system. Though he knew of shortcomings in the *sexual* system, he felt it should be accepted universally and be improved piecemeal, for "to disturb it then would be unfortunate." But within a few decades the superiority of the *natural* system was apparent to all, and a rapid conversion of most American botanists to it occurred after 1831 when John Torrey published the first American edition of Lindley's *Introduction to the Natural System of Botany.* The debate marked the transition of science from the realm of the amateur to that of the professional. With burgeoning knowledge of increasing complexity, the specialist who discussed recently acquired knowledge and concepts in new, arcane terms left the part-time dabbler (merchants, physicians, clergymen, and lawyers) feeling impatient and resentful.

Despite Rafinesque's strident advocacy of the *natural* system, he seemed to appreciate the convenience and relative ease of use of the Linnaean system of classification, especially for amateurs. In 1827 he published *An Essay on Botany* in the Saturday Evening Post in which he expounded on "this amiable science peculiarly suited to afford amusement to the ladies in the country." On Linnaean principles, the plant kingdom was divided into twenty-four classes, based largely on the number and nature of the stamens and pistil of the flower. Cryptogram ("ferns, mosses, seaweeds, fungies, and mushrooms"), where the "flowers are not perceivable to the naked eye" constituted the twenty-fourth class.[29]

With an understanding of the rules clear to him, Rafinesque went about "rectifying" approximately five hundred existing genera that he considered "preposterous or artificial," creating in their place "natural and proper" genera. He justified his high-handed correction of names other botanists had assigned (much to the chagrin of the original authors), claiming that they had not understood the rules and definition of *genera* and *species*; therefore, their assignments were based on false premises and were incorrect. The naming of species was less important "since they are variable," but still they must be rigorously identified, their "fixed forms" along with their variations. Presciently, he believed that varieties may ultimately assume "specific rank by important features, united to permanency" upon isolation in distinct climates and that species are productions within genera that started out as a single type.

Of interest is one of the examples he dwells upon—mankind. "The genus HOMO, once a single TYPE, that has produced during many ages so many natural varieties and breeds," the naming of which, whether they are called "species or races, Breeds, or Proles, Varieties . . . is immaterial." In matters of race (in the modern sense), Rafinesque was on the side of the angels, most certainly an abolitionist, determined not to provide arguments to bolster prevalent racist attitudes in the United States.[30] Although Rafinesque wrote of competition in the plant world, with one kind of plant replacing another in the wild though not in an evolutionary context, competition between animal populations was never discussed, perhaps because it could provide a "natural," biological justification for the abusive dominance of white Europeans over black people and Native Americans.

Believing that botanists were burdened with too many "useless" names that had no real basis in nature, Rafinesque sometimes reclassified plants and from time to time casually broke rules of nomenclature. If a specific or generic name (assigned by others) was too short, he would take it upon himself to lengthen it, and if too long he would shorten it. He preferred Greek names but disliked names that were a combination of Greek and Latin ("mongrels"), and he would "correct" them; others that he considered "uncouth" or "harsh" he would improve—a service that did not endear him to the original authors. Imperiously, he sometimes formulated new rules of classification as he saw the need and created and published new names, based solely on the descriptions of others, in his reviews of their work. If he could detect a difference between the characteristics of a plant presented in a publication and his own conception of that plant, he would rename it!

His astonishing ability to invent and name new sciences became absurd, and fueled his enemies' ire—Socapology, Rytology, Leptology, Gazaplogy, Sercology, Phlegology, Metallogy, Socadology, Gazology, Anapatology, Atmisology, Ychrology, Sycreology, Thermiology, Eleiology, Sphaltology, Coniology, Oxydology, and Aiology just to name a few—and several of his colleagues had to admit that his names were seemly, euphonious, and meaningful.[31]

On two occasions Rafinesque attempted to name genera after himself, *Rafinesquia,* but both failed because the genera were reclassified with new names. Today there exists a plant genus *Rafinesqui* (in the *Compositae* family) that contains two species. There are also several species—a cactus, brachiopod fossil, coral, bat, and fish, bearing the specific name *Rafinesqui,* a name that was bestowed by others to honor him.

Like a zealous missionary, Rafinesque suffered for his beliefs and his ideas, for he did not hesitate to express them in cautious, unreceptive company. American botanists berated him for advocating the new French system, and it is quite probable that Rafinesque made enemies by forcefully defending his approach. Wounded, he would respond with his usual tactlessness and widen the breach. In a letter to Bory St. Vincent that he published in Lexington, Kentucky, while living there, he had this to say about some of his fellow townspeople—his neighbors and colleagues:

> A set of unfortunate individuals, who have two eyes; but cannot see: their minds are deprived of the sense of perception: they are astonished and amazed at my discoveries, are inclined to put them in doubt and even to scoff at them. The art of distinction is entirely unknown to them; they are like the uncivilized savages who call cabins all our various buildings, let them be huts, cottages, log-houses, brick-houses, stone-houses, barns, churches, palaces. . . . Thus our cat-fishes, eels, shads, sturgeons, &c. are for them mere fish to fill their stomach! and moreover they are all of European breed, and were carried here by Noah's flood direct from the Thames, the Seine and the Rhine! —I let them rail to their heart's content, and I laugh at them. . . . It is only in Europe that my labors and discoveries may be fully appreciated: here I am like *Bacon* and *Galileo,* somewhat ahead of my age and my neighbors.[32]

After unburdening himself of these acidities, Rafinesque did not consider the situation hopeless, reasoning that by adding a few kind words, no one would take offence. "I am however happy to perceive that this apathy and reluctance for scientific researches is very far from being general: we have already at this early period of existence of these western states, [Kentucky and Ohio] as many enlightened citizens and writers as in any part of Poland and Russia of equal extent, already more than in our southern states, and will soon rival and surpass the middle or eastern states."

This young, odd-looking foreigner, articulate but without small talk, did not fit easily into a society of well-to-do, self-satisfied citizens. Rafinesque was an enlightened theist who must have frequently encountered American fundamentalism and biblical literalism. His scathing, outspoken remarks about the origin of New World fish, or on the "absurd" notion that Indians were one of the lost tribes of Israel could not help alienating the religious community, to which many naturalists and botanists belonged.

Rafinesque was a field botanist, an acute observer of plants in the wild, and he had a fine appreciation of natural affinities among different species. He was, in fact, considered by some to be the best field-man of his time in America. In the wild, the forms of plants might vary according to inheritance (of which almost nothing was known), and environmental conditions—geographically isolated groups of plants might begin to deviate from parental populations. The closet naturalist in his sheltered workplace who classified a limited number of plants from the field strove to idealize the form of the species before him, to minimize or even ignore variations within species, and to lump variants together to create a tidy system with a minimum of internal exceptions and inconsistencies that would only confound the classification process. Variation troubled other natural historians such as Thomas Say, who noted the extreme variability of beetles, but Say had even less insight into the meaning of variation than did Rafinesque.

Workers in the field, like Rafinesque, were to be pitied for they were often challenged and overwhelmed by the variations with which they were confronted. When deviation from the idealized type was sufficiently pronounced, and this often occurred, field botanists would create a new species. Disdainfully, these botanists were called "splitters," people who alarmed "lumpers" by threatening to inundate them with so many new genera and species that the entire classification system would be placed in jeopardy.[33] Rafinesque felt justified in his creation of new genera and species, for with an acute, experienced eye he rightly concluded that many plants had been incorrectly forced into taxa by arbitrarily ignoring differences in certain important characteristics, where new groupings would have been more appropriate. In other words, taxonomists were confronted by a troublesome question; if a species, which by definition had unique characteristics, was found to have two (or more) newly observed versions of one of its "unique" characters, should the original species be elevated to the level of a genus that contains the two (or more) new species? By delving further and deeper into an enlarging circle of characteristics, an ever-increasing number of species could be created—a species splitting process. Furthermore, with the subtle variations in plants constantly created by environmental influences, a whole continuum of differences might overwhelm the taxonomic process. Thus, for practical reasons, the number of members of a taxon has always been a concern of taxonomists, for if a taxon grows too large, human memory falters and the system becomes too unwieldy.[34] In the end, reliable classification necessitates making judgments on several specimens that form a reasonably homogeneous group, and the deviations of the determining characteristics must be within a statistically acceptable range.[35]

The supreme splitter of American and even of world botany, Rafinesque gave names to approximately 2,700 new genera, 320 new subgenera, 6,700 new species, and nine hundred new varietals. However, the overwhelming majority of these did not stand the test of further analysis. Linnaeus himself comes in a poor second with 1,500 genera ascribed to him, many simply taken over from others.[36] In the long run, for the stability and order of the classification system, it is fortunate that Rafinesque's numerous new names, usually untestable and obfuscating, were simply ignored by American botanists with their intuitive good sense. Heightening their general dismay was the poor editing of many of Rafinesque's publications, the inconsistency of his terminology, errors in his spelling, and his overly brief definitions—all having important consequences.[37] Since the use of plants and plant products was of major importance in medical practice at that time, the casual change of the name of a plant could confuse the compilers of pharmacopoeias and physicians, with potentially serious outcomes.

To Rafinesque, the validity of his work lay in his firsthand experience with nature, epitomized by the statement: "I shall describe here only those [whole molluscs, not just the shells] which *I have now before my eyes.*"[38] Yet he frequently contradicted himself, for he published many descriptions of plants and animals he had never seen. He had written a book on the plants of Louisiana based upon descriptions of an amateur botanist, without ever seeing the plants or Louisiana—an egregious practice that was widely discussed. After berating several American conchologists by name (Say, Lea, Eaton, Barnes) for ignoring his published work and being "led astray by various motives," he singled out Thomas Say: "Mr. Say is above all, inexcusable." Apparently, Say had made a hash of some of his descriptions by concentrating on the shell of the mollusk rather than the animal inside and in doing so mistook the mouth for the tail. This blunder aroused in Rafinesque a fury worthy of an Old Testament prophet: "If he had seen these animals alive, feeding, moving, and watched their habits as I have done repeatedly, he would not have fallen into such a blunder. . . . Others pretend that my monograph is too intricate; it is the subject which is such whenever many species belong to a tribe, many divisions and sections are needed to elucidate and isolate the species. All the great naturalists know and do this."[39]

Closet botanists, the elite "thinkers" of the field, were able to consult books and reference collections for assistance, but field-workers did not have this luxury, working under less than ideal conditions, too close to the objects of study to be objective, too immersed to keep their perspective. In the wild, at the frontier of the country, they were usually not as well read, nor had they access to up-to-date reviews of the relevant literature, as did workers who led an orderly life in a large city with libraries and universities at hand. Rafinesque must have had a remarkable memory to keep thousands of names and taxonomic relationships in his head while wandering over the land, with little access to books. Since he was in the field much of the time, far more than any of his peers, his errors and lapses are understandable. Often, what he seemed to publish were his rough, incomplete descriptive field notes, perhaps because he was in too much of a hurry to review them thoughtfully, and properly amend the jottings he had made while in the field. It is known that in some instances Rafinesque made no notes while on summer expeditions but wrote up his work from memory the next winter. This sort of loose practice may partly explain why Rafinesque was accused of not giving due credit to predecessors and colleagues and of sometimes giving two different names to the same plant.

Charlotte M. Porter has pointed out that field-workers like Rafinesque, John Kirk Townsend, Thomas Nuttall, Thomas Say, and Titian Ramsay Peale were effectively "suppressed" by the closet naturalists who were mainly of the conservative establishment. All suffered from a lack of funds, and some were marginalized for reasons of political radicalism and unconventional lifestyles.[40] Field-workers were not out of place in the field clubs of amateurs and in their networks. It was the tidy closet naturalist, the pioneering professional botanists with institutional support, the wealthy merchants and physicians leading stable urban lives, who after their adoption of the natural system brought a new, critical level of performance to taxonomy.

Botanical information was systematized and incorporated into an index, and as information on new species of plants came along it was inserted into supplements. An international inventory of all plants such as *Index Kewensis* (associated with Kew Gardens, London) could be consulted to see whether a plant had been previously described. Through them a body of detailed knowledge grew that left the dabbling amateur far behind. A full appreciation of any branch of science increasingly required a specialized education. Closet naturalists also founded professional societies, and controlled the economics of science, such as they were, as well as access to publication.[41] Publication in reputable journals could easily be prevented by the peer review system or by the simple misplacing of manuscripts. Forced to publish at their own expense, the papers of field-workers were criticized for their poor technical quality. George Ord, an influential naturalist in Philadelphia commented acerbically on the language and the quality of one such publication: "I am sorry to be compelled to add, that the paper and letter-press are a disgrace to the arts of our country. A book possessing such repulsive characters could hardly hope for general encouragement."[42] According to this scornful judgment, Rafinesque's publications were beneath contempt, for many were badly printed on such poor paper that they disintegrated within a few years.

There is little doubt that serious criticism of Rafinesque was justified. Thomas Say, a man of irreproachable integrity, who had also suffered at the hands of the establishment, was infuriated by Rafinesque and was in a position to let everyone know the reason for his outrage. When he was a conscientious editor of the *Journal of the Academy of Natural Sciences of Philadelphia*, he accepted a paper from Rafinesque on a fish, only to find that he had previously published it elsewhere with a different name for the fish described. Some said Rafinesque's craving for priority and fame were responsible, but disorganization and forgetfulness may have played an inglorious part. The paper was retracted, and thereafter, Say refused to accept any article submitted to the Journal by Rafinesque—the first articles ever to have been rejected by the Journal! Since Benjamin Silliman, editor of the *Journal of Science and the Arts,* also rejected Rafinesque's papers, by 1819 the two major American scientific journals were closed to him.[43]

Rafinesque was Say's bête noire. As the professional zoologist on the historic Long expedition to the Rockies in 1819 and 1820, Say had named several new, important species, only to find the names were invalid because according to the rule of priority, the animals had been identified and named previously by the accursed Rafinesque. Say had come across the mule deer and had named it *Cervus macrotis* only to find that Rafinesque had previously named it *Cervus hemionus;* the mud puppy, a salamander with external gills he had named *Triton lateralis,* but this was disallowed because Rafinesque had named it *Necturus maculosus* the previous year; Say had created a new genus *Pseudostoma* in which he had placed the plains pocket gopher, but Rafinesque again had previously assigned it to the genus *Geomys;* similarly, Say named the prairie rattlesnake *Crotalus confluent,* which he later learned had been previously named *Crotalus viridis.*[44]

As an exceptional observer in the field, Rafinesque seemed to be more aware of biological variation than almost anyone. Rather than ignore this troublesome phenomenon, he proposed a brilliant explanation for it briefly in 1814, and more fully in 1832, twenty-seven years before Darwin's *On the Origin of Species.* This was a time when most people accepted without question the notion that genera were immutable and species were fixed, a conception that precluded any evolutionary process. The notion of fixity is ancient, originating in Plato's ideal forms or "essences," with variation being nothing but illusion and of no consequence. Aristotle introduced the idea to biology, and since that time, the notion has had numerous, prominent advocates, including Louis Agassiz. Darwin himself believed in the immutability of species in 1828 when he sailed on the *H.M.S. Beagle*, and it was not until the mid-1830s, and especially after 1838 when he read Malthus's *Essay on the principle of population* that he began to formulate his ideas about variability, natural selection, and the origin of species.[45] In 1832 Rafinesque had written:

> The truth is that *Species and perhaps Genera also, are forming in organized beings* by gradual deviations of shapes, forms and organs, taking place in the lapse of time. There is a tendency to deviations and mutations through plants and animals by gradual steps at remote irregular periods. This is a part of the great universal law of PERPETUAL MUTABILITY in everything . . . every variety is a deviation which becomes a Sp. as soon as it is permanent by reproduction. Deviations in essential organs may thus gradually become N.G. [new genera]. Yet every deviation in form ought to have a peculiar name. . . . It is not impossible to ascertain the primitive Sp. that have produced all the actual; many means exist to ascertain it: history, locality, abundance, &c. This view of the subject will settle botany and zoology in a new way and greatly simplify these sciences. The races, breeds, or varieties of men, dogs, roses, apples, wheat, . . . and almost every other genus may be referred to one or a few primitive Sp. yet admit of several actual Sp. names may and will multiply as they do in geography and history by time and changes, but they will be reducible to a better classification by a kind of genealogical order or tables.[46]

Again, in his *New Flora of North America* (1836), he restated the process luminously, calling it one of the "dark mysteries of generations past": "Every species was once a variety, and every variety is the embryo of a new species." Rafinesque was describing an evolutionary tree, with members of the tree related, the consequences of macrochanges that resulted in the formation of new genera and microchanges that operated at the species level.

To Rafinesque's peers, wild speculation such as this was evidence of his madness. Darwin himself, reviewing the history of transformism and species formation, mentions Rafinesque for his exposition of the idea.[47] Whether Rafinesque held this subtle and prophetic notion at the time of his first visit to America in 1802 to 1805 is not really known. He probably did, for he was acutely aware of small differences and was splitting species and genera from the very beginning of his career, a practice that was completely consistent with his views on the formation of new categories of plants and animals. But nowhere in his discussion was there mention of the possible mechanism or benefit of variation, which remains Darwin's (and A. R. Wallace's) preeminent contribution—the theory of natural selection. Nor did he comment on Lamarck's notions of the heritable effects of chronic use and disuse of parts as a driving force for change. A developed, thoughtful discussion of his views on classification and evolution is absent from Rafinesque's writings. His notions were probably formulated early in his career, because some credit for the idea of change belonged to Michel Adanson,[48] an important early influence on Rafinesque. He was also aware of the work of Lamarck, Buffon, Erasmus, Darwin, and Aristotle, all of whom had touched upon transformations, changes, in living forms.

The notion of transformation and change was given poetic and philosophical voice in Rafinesque's *World, or Instability* (1836), a 207-page poem in twenty sections in the style of Milton in which he sang of a God of Love who presided over the infinite, unfathomable complexities of Nature and the Universe. Rafinesque became a prophet wandering in the American wilderness, donning the mantle of Moses, a supplicant conversing with God and writing a new Scripture:

> In endless shapes, mutations quick or slow,
> The world revolves, and all above, below,
> In various moulds and frames all things were cast,
> But none forever can endure or last.
> Whatever took a form, must change or mend;
> Whatever once began, must have an end.[49]

To staid American botanists Rafinesque's ideas were a threat in their novelty. They *knew* that the absolute stability of every species, each the perfect work of the Creator, was beyond question. The classification of species (and genera) was dependent on this static view, which was sanctified by Cuvier and by Agassiz. Rafinesque's proposal, "related species of such genera as *Rosa, Quercus, and Trifolium* have had a common origin," was emphatically rejected by his contemporaries, but it is now dogma. According to David Starr Jordan, (who along with many other nineteenth- and twentieth-century scientists had a "peculiar interest" in Rafinesque), Rafinesque sounded better in 1888 than he did to his contemporaries in the 1830s, who loathed and pitied him and his science.[50]

Rafinesque raised the troublesome problem of how one defines "species," and the arbitrariness of boundaries. Where does one species end and another begin? Can members of one species give rise to another species or genus? Serious dispute and challenge of fundamentals were unseemly to the establishment of gentlemanly, contented botanists and zoologists, and so Rafinesque's advanced ideas and insights, some without foundation, enhanced Rafinesque's reputation for unreliability. In 1841, Asa Gray, a future evolutionist, wrote a contemptuous obituary of Rafinesque in which he spoke of his bizarre notions: "According to his principles, this business of establishing new genera and species will be endless; for he insists, in his later works particularly, that both new species and new genera are continually produced by the deviation of existing forms, which at length give rise to new species."[51]

Rafinesque's rehabilitation began after the publication of Darwin's *On the Origin of Species* (1859), for as fundamentalism lost its strength, the frozen boundaries between taxa thawed, and the notion of trees of descent gained acceptance. Rafinesque's ideas made some sense once it was accepted that variation within species was constantly taking place and that species within a genus might have a common ancestor. Taxonomists later realized that other reputable naturalists of the early nineteenth century could be criticized for the same Rafinesquian shortcomings, born of the imperfect and confused state of taxonomic practice and knowledge of the time. The perspective of history has softened the severe condemnation of Rafinesque and has fueled the rehabilitation of his reputation.

Upon returning from one of his numerous botanical expeditions, Rafinesque found a letter awaiting him from Europe, perhaps containing an offer or some information and advice he felt he could not ignore. No material assistance seemed to have been proffered. Despite the fact that "several of my friends wished to detain me, and made me several offers of employment, not quite to my taste," the young man returned to Europe bearing his herbarium of ten thousand specimens, representing nearly two thousand species. His departure was not only prompted by his wanderlust, for he dreamt of exploring Greece and Asia, but it was becoming apparent that positions in science in America were scarce and the few that appeared would not be open to him. His ability to publish his ever-burgeoning writing was limited and perhaps even nonexistent. He must have realized that in order to continue doing what he liked best, he needed patronage, public or private, and this was not forthcoming. His reminiscences abound in references to his "friends," and he frequently mentions places where he was "well received" as if desperate for acceptance and affection—he was keeping score. As a man without resources who looked for advantage, it was important for him to know who his enemies were and where his opportunities lay. One senses that he made the rounds of important people who might help him, and for the most part he was treated with tolerant politeness, but was offered little.

The hard reality was, as Rafinesque and other worthy naturalists learned, there was almost no possibility of making a stable living in natural history and science, at a time when most Americans had to earn their bread as professionals, farmers, merchants, and businessmen with little time to spare for "frivolities." American institutions of higher learning, such as they were, were not capable of turning out well-trained scientists and naturalists, for the teaching of these disciplines was practically nonexistent, and if taught, a single professor was assigned the task. Young people attended university for instruction in the classics, theology, rhetoric, Latin, and Greek, usually to prepare themselves for medicine, law, or the ministry, or at the very least to develop their intellect. If science played any role at all, it was to illuminate the ways of the "Creator."

Rafinesque's first encounter with the New World had come to an end; his romance with America was over for the present, after having seen the land as few had ever seen it. He had become increasingly aware of hostility, and he tactlessly answered in kind, though it was particularly important for him to impress upon his audience of readers that his peers accepted him, both socially and scientifically. If Rafinesque had been born to wealth and had come from a prominent family, his story would have been radically different. As it was, so much of his effort was taken up with the problem of survival. Rebuffed, looking upon himself as an American, a bruised Rafinesque retreated to a familiar Mediterranean environment where he thought his brilliance would be appreciated. Ten years later Rafinesque returned to the United States, and five years after that he became an American citizen.[52]

Chapter 3

SICILY
1805–1815

The Rafinesque brothers boarded the ship, the *Two Sisters*, in late December 1804 bound for Leghorn (Livorno), Italy, but as they left Philadelphia heading for the open sea, an unusually early drop in temperature caused the Delaware River to freeze over in a matter of hours. After several weeks, cutting through miles of frozen river, they finally reached Delaware Bay, and they were on their way. A stormy crossing of the Atlantic took thirty days, Gibraltar was sighted, and six days later they reached Leghorn—only to be shipwrecked at the entrance to the harbor. To add to their misery, a forty-day quarantine was imposed on them "without cause." Rafinesque spent his time "arranging my plants (*10,000 specimens*), drawing the news species, writing my travels and letters." The quarantine was not taken too seriously, for unofficial arrangements were made that allowed him to see his mother, sister, and friends, and to exchange specimens with Italian botanists.[1]

In March, the twenty-two-year-old Rafinesque set foot on Italian soil, and while preparing for the continuation of his trip to Sicily, his final destination, he explored the area around Leghorn for new plants. By May he was off to Palermo, parting with his brother Antoine, who with his mother and sister returned to Marseille. Sailing south along the Italian shore to Corsica, Sardinia, and Utica, was a "happy" time for Rafinesque, but when the ship arrived at Palermo, again they were quarantined for twenty days because Yellow Fever had been reported in Leghorn, their port of origin. Never mind that the information was a year old!

At the time, Napoleon, now a self-proclaimed emperor, controlled almost all of the monarchies of Italy except for the Kingdom of Sicily, which remained under British military control, ruled by puppets—King Ferdinand and Queen Maria Carolina, formerly of the Kingdom of Naples and Sicily. Britain closed mainland ports by naval blockade so that Sicily could only trade with England, America, Spain, Barbary, Sardinia, Malta, and the Levant. The French were not popular in British Sicily, and they were increasingly disliked (for good reason) by the mainland Italians. Of the 20,000 Italian volunteers and conscriptees who had accompanied Napoleon to Russia, only 333 had returned. Over decades of turmoil, there were no satisfactory political alternatives for Italy until the Risorgimento, the unification movement, at midcentury. But anti-French sentiment was clearly taking shape and the name Rafinesque was very French, so Constantine prudently added his mother's very German name to his own. He was now Constantine Samuel Rafinesque Schmaltz, passing as an American who wrote in Italian rather than in French. Rafinesque Schmaltz prudently published *Specchio delle Scienze* in Italian, though many of his correspondents were in France. Surprisingly, he did not publish in English, for his new orientation was toward the United States and his close scientific colleague William Swainson was an Englishman working for the British army in Sicily. Perhaps with his dim American prospects and a future that was uncertain, Italy seemed to him to be a reasonable destination for his talents, and his Italian writings would, he hoped, give him status or at least garner membership in Italian or Neapolitan Academies that could adorn his name.

Rafinesque Schmaltz, a child of the Mediterranean, was enchanted with Sicily:

> My first impressions of this lovely island were delightful: arriving in the month of May, the air was embalmed by the emanations of orange blossoms, carried far at sea in the night by the land breeze. The mountains were smiling with flowers and verdure, they invited me to climb over them. . . . Here I was then in Sicily the largest and finest of the Islands in the Mediterranean: a residence of ten years made me perfectly acquainted with it and its natural productions. Few learned travellers can boast to have so long studied Nature in that lovely spot. It was the best epoch of my life. The events of those ten years might afford materials for a romance. . . . Sicily might be described in a few words by saying that she offers . . . a fruitful soil, delightful climate, excellent productions, perfidious men, deceitful women.[2]

He had no trouble with nature, he mastered nature, and he could freely rhapsodize about flora, fauna, and place, but it was his experiences with humans that usually ended badly.

Though not yet an American citizen, he wrapped himself in its flag, becoming Secretary to Mr. A. Gibbs, the U.S. consul in Sicily, who was a banker and merchant, and he signed his letters "Chancellor of the American Consulate." With his command of French, Italian, English, and other languages, and with America's and his own interests uppermost, his position at the Consulate at Palermo was privileged and important, and it left him with free time to indulge in his own business affairs. He lived with the Consul in a palace, and by 1807 he had made a "small fortune" that permitted him to have a house of his own in Palermo and to resign his official position. The parting appeared to be amicable, for Rafinesque arranged for his brother Antoine to replace him in the Consulate. While Constantine roamed the island, Antoine stayed close to home, tending to business, and he continued to do so for a few years after Rafinesque had left Sicily.

Despite the decline of trade in Sicily during wartime, Rafinesque, ever the entrepreneur, made his "first personal fortune" as a trader of natural products of medicinal value derived from regional plants, some of which he had discovered. By January 1806 his business, which exploited the plants of the island and the surrounding sea, was thriving. Incomprehensible to the Sicilians, he organized the large-scale processing of the familiar and "useless" squill, the bulb of the Sea Onion. There was no shortage of them, nor of laborers to gather them. Sliced and dried bulbs were sold to Americans, Russians, and the English for use as a cardiotonic, a diuretic, and an expectorant and for the making of a special vinegar; combined with paregoric it was widely administered to children, who were made miserable by its use. Rafinesque bought fresh bulbs for one dollar per hundred pounds and sold the dried product for ten to thirty dollars. In a few years, before Sicilians really understood what he was doing, he had prepared and sold 200,000 pounds of squill. He also developed a trade in rosemary, wormwood, and bay leaves. Belatedly, Sicilians discovered there was money to be made selling natural medicinal and herbal products, and so they set up rival manufactories.

His business ventures extended beyond the medicinal, for he wrote of traveling seventy miles to Cefalu and Tusa to oversee the loading of "oil" onto an American ship, and in another commercial trip he rode in a carriage to Termini, twenty-four miles from Palermo, to load two ships with wheat. He also mentioned purchasing Barilla, an alkali-rich sea plant used in the making of soda, soap, and glass. In 1813, acting for a "society of gentlemen," he set up a brandy distillery at Misilmeri that employed superior technology and design. With perfect aplomb he makes the remarkable statement: "I made a very good Brandy, equal to that of Cette and Spain, without ever tasting a drop of it, since I hate all strong liquors," but after two months he abandoned the venture because it no longer interested him.

Rafinesque never let his commercial inclinations dominate his life for long.

As his interest in commerce waned, his appetite for the natural world reemerged, and he became restless and began to roam again. Initially, he visited the towns and countryside near Palermo and then went further afield to nearby mountains and villages famed for their palaces and gardens. As in America, he befriended and collected cultured, influential people—naturalists, conchologists, botanists, and even an astronomer of the region—just as he collected fossils and minerals. He also developed a special interest in fish and molluscs, classifying and drawing them.

Sicily is a bridge between Europe and Africa, where plants native to both land masses flourish and provide a seedbed for Grecian and Spanish plants. Relatively little work had been done on the flora and fauna of the island and much of what had been done was, according to Rafinesque, of poor quality. From his point of view, this was virgin territory—his territory—and he exulted when he came across a rare or undescribed plant. Indeed, nothing in the world gave him more pleasure. The richness of the surrounding sea was exceptional—Rafinesque wrote about four hundred species of fish, "shells, molusca, zoophytes and sea plants without number." These he had small boys collect for him along the beaches, and he regularly canvassed fishermen for unusual specimens. He visited the ancient ruins of Greek temples at Segesta and Agrigento; inspected sulfur mines, quarries, and volcanoes; and rambled over mountains, valleys, and deserts, observing and collecting. The more remote and difficult the terrain, the rarer the plant he would find. Pennell, a botanical authority, wrote of the publications of that period: "We may fairly complain of the brief and sketchy nature of the descriptions of new species, but they testify to Rafinesque's keen observation of our flora, and most of his new names have been received into all our books."[3]

Since Sicily was under the protection of Britain in wartime, Rafinesque was unable to keep in touch with Paris, a serious isolation from his French peers at an early time in his career. However, he was in contact with some of his colleagues in America, but distances were great and communication was difficult. Rafinesque had written a summary of all the genera and species of North American plants he had come upon and named during his visit. Unable to publish the report in France, he shortened it and sent it to his friend Samuel Latham Mitchill in New York. Mitchill, editor of the widely read journal *The Medical Repository*, published the report as a series of articles and excerpts of letters.[4]

In addition to the botanical supplement, Rafinesque wrote of quarantines in Sicily, French procedures for fumigation, plants of medicinal importance, and European plants naturalized in the United States. His enthusiasm about

American plants mounted, and with little sense of the possible, he announced his intention to write a comprehensive account of North American plants, especially those he had discovered. The treatise was to be accompanied by a companion work on the fungi, all to be done while in remote Sicily! Surely Mitchill must have read this announcement with amazement and perhaps with some amusement (or had the very busy Dr. Mitchill read it at all?). One cannot but wonder why he published this impossible prospectus.

The first of Rafinesque's Italian publications, *Caratteri di alcuni nuove generi e nuove specie di animali e piante della Sicilia* (1810), and *Indice D'Ittiologia* were concerned mainly with fish, some entirely new, which anticipated the work of Cuvier. This the great man acknowledged, but he did criticize Rafinesque for describing "imaginary" fish. These was the first written reports on Sicilian fish, and in his various Sicilian papers Rafinesque described many hundreds, even thousands, of new species and genera of fish, bats, birds, amphibians, crustacea, molluscs and insects. In Rafinesque's obituary, Haldeman derogated much of this work; while he admitted that one work from 1810 was "good," he damned the rest. "His greatest fault was not so much, perhaps, the shortness and resulting obscurity of his characters, as his passion for 'new species,' and the recklessness with which he proposed them, without sufficiently examining the work of his predecessors," and he went on using such derogatory terms as "not new . . . doubtful . . . imaginary . . . reckless."[5] There was considerable justification for criticizing Rafinesque, but Haldeman and other members of the Eastern establishment, such as botanist Asa Gray,[6] and conchologist Isaac Lea, were excessively exacting about his Sicilian and subsequent writings.[7]

On the other hand, biologist William Swainson (1789–1855), Rafinesque's friend and companion in Sicily, vigorously defended his work, claiming that it was never fully appreciated. He had seen Rafinesque making extensive notes on fresh, sometimes living specimens, quite different from the imperfect, discolored specimens used by Cuvier that he had received from Rafinesque. The specimens had deteriorated during shipment, something Rafinesque could not prevent, but Cuvier was annoyed and had blamed Rafinesque for this deficiency. Swainson dismissed Cuvier's criticisms, attesting to the accuracy of Rafinesque's observations and the quality of his descriptions and drawings from which the fish depicted could be easily identified.[8]

Rafinesque had met the English botanist, ichthyologist, and naturalist William Swainson at Messina where the British army commissariat was encamped, and over the next few years the two men did fieldwork together during Swainson's posting in Sicily. "Swainson went often with me in the mts., he carried a butterfly net to catch insects and was taken for a crazy man or wizard. As he hardly spoke Italian I had once to save him from being stoned out of a field, where he was thought to seek for treasure buried by the Greeks."[9] Many years later in 1840, the year of Rafinesque's death, Swainson published a collection of short biographies of zoologists and botanists, which included that of his old friend, Rafinesque. He spoke of the pleasure of knowing him, praised him for being a "most enthusiastic and persevering naturalist," but he understood the basis for his widespread rejection. Although he "discovered and described a great number of new objects," he judged his work to be almost useless because of his uncontrollable species splitting and the extreme brevity and carelessness of his descriptions. Because of these "vicious defects," he remains unappreciated.[10]

A consequence of Swainson's cordiality was that Rafinesque never knew Swainson's real opinion of him and his work. In 1828, replying to a letter of George Harlan, who had recently had a bitter dispute with Rafinesque, Swainson wrote: "Your account of poor Rafinesque, I am sorry to say, is perfectly just . . . we are intimately acquainted having both resided at Palermo, in Sicily, from 1810–1813 where he marr'd every subject he took in hand [the period when he was defending Rafinesque's work on fish]. Worse perhaps, than he has done in America! In that, his writings must never be depended upon"—a damning statement from a friend. Assuming that Swainson was not a duplicitous, opportunistic man, there was an ambivalence in his regard for Rafinesque that probably reflected the attitude of many of his colleagues. He was fond of Rafinesque with all of his enthusiasm and brilliance, and he could not bear to hurt this vulnerable man, yet he recognized his flaws. Significantly, the praise and defense were public and available to Rafinesque, while the damning criticism was private.

From 1809 to 1840 Swainson received many letters from Rafinesque, each sounding as if it had come from a kindly but hungry man who could not repress his hortatory impulses; each letter demanding a commitment. He asked for favors and wanted to know the latest news, and in return he offered Swainson specimens of all kinds, opinions, and much new information. Although he "wished to serve" Swainson, a letter from Rafinesque entailed many hours of labor to fulfill his requests—to find a position for him, to get him elected to associate membership in the Linnean Society, and to sell specimens for him. He was grateful for a good review of his work, "in the midst of injustices which has often overwhelmed me." Swainson remained a friend and became a confidant: "I burn of the desire of having the honor of being a Professor," and "I would go anywhere for a trifling annuity." Swainson became a wailing wall. Rafinesque wrote of his "dreadful fever" that kept him in bed for a month, his shipwreck in 1815, and he fumed at his enemies: "They feel my superior genius and want to crush me, as it has been done for 20 years past." In another letter he wrote: "Some petty American Naturalists, Lea, Barnes, Say are or were jealous of me, but they drop around me." Sometimes in despair, he demanded justice, and in the end he realized that "I depend on me alone"—utter dislocation and loneliness.[11]

In the summer of 1809, Rafinesque made a grand tour of eastern Sicily on mule, with servants attending to menial matters. When needed, local guides were picked up along the way allowing Rafinesque to study the geology of the area, sketch, and gather specimens of minerals and "a rich harvest" of new plants. The expedition culminated in his witnessing the rising sun from the lip of the crater of a smoking Mt. Etna, a "sublime" panorama, "the whole of Sicily at my feet . . . the highest mts of Sicily appeared now as mere mole hills!" Peering into the steaming cauldron he was awed, enchanted by its beauty, but terrified. He seems to have had a particular interest in volcanoes, lava flows, and volcanic rocks, commenting in passing: "I had already surmised the true theory of volcanoes, which Etna, the Azores and N. America have confirmed." However, what Rafinesque's theory of volcanoes was remains unknown.

Rafinesque's interest in volcanoes and geology is understandable in the light of the great controversy about the origin of the earth that took place during the first decade of the nineteenth century. There were two actively contesting schools of thought on the matter. One group, the Neptunists, believed that the earth had been completely covered by water and that all the surface components of the earth (rocks, minerals, etc) precipitated out, layer by layer, during five great stages. A second group, the Plutonists believed that the surface of the earth was fashioned by heat and pressure in the earth's core, which lifted up the earth's crust, often irregularly, to shape the earth's surface. Upheaval was followed by a constant wearing down of surface features by wind, rain and ice—a continuous process without beginning or end. In Rafinesque's time, it was believed that the dispute might be settled by a close study of volcanoes and the rocks within and around them, and so geologists and naturalists traveled the world studying volcanoes. Since Rafinesque was in Sicily, an island with active volcanoes, he seized the opportunity to study the volcanic Mt. Etna.[12]

The expedition was not without danger. Despite the presence of murderous bandits patrolling the hills, a foolhardy Rafinesque deemed himself safe from them because he was a "botanical physician" who surely would not be molested by any living soul—"I never met any in a band, and they seldom attack those they know without money and weapons to defend it." His questionable reasoning turned out to be accurate; unarmed, he encountered a heavily armed robber who treated him rather well, regaling Rafinesque with his own tales of adventure, or "pranks" as he referred to them. All this thief wanted was corn and cheese, and the best bed in the cottage in which Rafinesque had taken shelter. Wolves and foxes surrounded them during the night taking sheep and goats despite the shepherds' dogs, and they slept with the howl of the wind in their ears. Rafinesque continued the next day unharmed.[13]

Upon his return to his home in Palermo, bringing with him a "beautiful collection of minerals, volcanic objects and plants," Rafinesque was laid low by a "malignant bilious fever," from which he slowly recovered. "It is almost the only serious malady that I have experienced through my excursions, thanks to a good constitution, and a rigid sobriety."[14] The experience must have been very serious because it dampened his ability, if not his ardor, for travel over the next few years. Instead, he devoted more time to writing. He continued to publish lists of his botanical and zoological "discoveries," and with Giuseppe Emmanuele Ortolani, a Sicilian mineralogist and lawyer, he embarked upon a geographical and statistical description of Sicily that included political, constitutional, and commercial information—an impressive work that was printed at Rafinesque's expense. Unfortunately it was censored, in part because it could have been of use to the French who were suspected of preparing to invade Sicily.[15]

Despite Rafinesque's enchantment with Sicily he was "disgusted with Sicilians," and was eager to leave the island as early as 1810, finding it as difficult to find a position in the Old World as it was in the New. In 1812 he applied for the chair of botany at the University of Palermo, only to find that the position had secretly gone to the son of the former, deceased professor. Two years later the chair of agriculture and economy position opened and was filled by a government minister's clerk. Rafinesque was embittered by the subterfuge, for both positions were supposed to be determined after an open competition, which he was confident he would win. Casting about, he established a correspondence with Sir Joseph Banks, the famous and wealthy president of the Linnean Society, who as a young man had served as botanist on one of Captain Cook's expeditions to the South Seas. Rafinesque offered to explore the shores and rivers of Australia for the Society, but Sir Joseph not only declined the offer, he also did not oblige Rafinesque and publish his monographs.

Early in his career his powerful instinct to express himself in publications hadn't fully come into play, and as a mere twenty year old, he certainly did not have the means to have his work printed. His writing and publishing mania first found expression in an explosion of print toward the end of his stay in Sicily. Having made some money, he was able to build an extensive library, and he could now afford to establish, edit, and support his own "encyclopedic" journal.

Rafinesque's journal, his first, was marred by every deplorable trait for which he was known throughout his life. Grandly he called the journal *Specchia delle Scienze*, the *Mirror of Science* or *Encyclopaedic Journal of Sicily*, *Literary Repository of Modern Knowledge, Discoveries and Observations on the Sciences and Arts and Especially on Physics, Chemistry, Natural History, Botany, Agriculture, Medicine, Commerce, Legislation, Education etc.* Though titles at that time were rather imposing and pretentious, Rafinesque's outdid them all, and one cannot fail to detect his determined hucksterism. By necessity he was editor, proprietor, and sole contributor. Indeed, he was one of the very first, if not the first to father such an enterprise. Every article in the *Specchia*—over two hundred—was written by him in the course of one year. While there were only two minor papers on archaeology and one on ancient history, he expounded on botany and zoology, meteorology, geology, geography, agriculture, economics, medicine, materia medica, pharmacy, chemistry, physics, mineralogy, metaphysics, legislation, commerce, statistics, the arts, architecture, and prison reform (especially reform of American prisons in Philadelphia and New York)—a virtuoso performance. However plants and animals always remained the special concern of most of his papers.

The journal was published as a monthly but, in fact, appeared at irregular intervals. It began with a prospectus—its aims, its amazing breadth, and the hope that it would have a worldwide influence—but like Rafinesque's later journals, it appeared in small numbers and lasted only one year because there was no demand for it. How could Rafinesque avoid the obvious conclusion that there were so few who were interested in what he had to say and that he was isolated geographically and intellectually? Driven, feeling much abused, he could not stop observing, analyzing, and writing about his master scheme for "the methodological study of natural history."

Rafinesque's idol, Linnaeus, was the creator of one such scheme that encompassed the physical world. He had divided the "productions of nature" into two empires—the mineral (inorganic) and the organic—and each of these empires was divided into two kingdoms. The mineral empire was divided into the elementary and the fossil kingdoms, while the organic one was divided into the animal (zoological) and the vegetable (botanical) kingdoms, each of which had ten classes. Classes were divided and subdivided to form elaborate schemes with many new, invented names, and it is here in this categorizing and naming process that Rafinesque felt he could make his mark, as he was ingenious in generating new names that were perhaps superior to Linnaeus in this regard. He created a new discipline of *Somiology*, the science of living bodies in which the general principles of nomenclature and classification were listed. The term was created by him, and it died with him. Then he announced that he had devised a greatly improved system of classification based upon the natural method, which he considered the "true and perfect natural method sought after by all Naturalists."

Over the course of several years in Sicily, he devised several systems, each an elaboration or improvement of the previous one; the fifth he called "synoptic." After stating that this ultimate system should compare favorably with those constructed by Linnaeus, Cuvier, Lamarck, and others, he humbly asserted "I dare to flatter myself that [I can] demonstrate its superiority," then ended by admitting that he has been perfecting his system for five years and that much work remained to be done. His mind ablaze, Rafinesque was too impatient to provide the details that would justify his enormous claims. His descriptions were cursory and often obscure, so it was difficult, if not impossible, to follow the trail of his thought. He simply could not resist trumpeting his grandiose plans for his unpublished work.

He was a born editorial writer in a land of tight control, who crossed the line of permissibility despite the undoubted loosening of stringent censorship by local authorities. His journal "succeeded well, but drew upon me persecutions and displeasures," although he was certainly not a revolutionary. He also became embroiled in an unresolved legal dispute with his printer, who refused to distribute the journal, though he had been paid. Sadly, his classification schemes were totally ignored by his fellow botanists, and indeed, Rafinesque himself did not in practice make use of his grand taxonomic edifice.

Two other publications in 1814, *Epitome of the Somiological or Zoological and Botanical Discoveries*, and *Fundamental Principles of Somiology*, were rehashes and elaborations of his *Mirror of Science* with some new data added.[16] In the latter work, he forcefully asserted the need for establishing a standard nomenclature and a system of classification, and to this end he listed over one hundred "laws or principles" in the manner of a Mosaic lawgiver. Among the many dedications of his *Principles of Somiology*, he extolled some of his colleagues: "to you also illustrious Buffon, Lacépede, Sonnini, Virey, etc., whom I honour in spite of your errors and your carelessness; your untamed Geniuses could not submit to rigorous rules [which his work will provide], while your writings make us understand at every step their necessity." Such ill-considered words, which could only inspire contempt and outrage and add to his already formidable problems, strongly suggest that he was manic and out of touch with reality; his delusions nurtured by his vast knowledge and abilities.

A year later he published his *Analysis of Nature or Tableau of the Universe and of Organized Bodies,* an ambitious summary of his cosmological and taxonomic ordering of all things in the universe, organic and inorganic (a synoptic view), and a further repository of data—a masterwork strongly influenced by Continental and French thinkers—for which he would want to be remembered. Befitting the weight of the enterprise, he declared that it was "a profound Study of the works of the Creator. . . . I have read in the great book of Nature: happily guided by the wise Linnaean precepts, it is in the dark forests of America and on the fertile strands of Sicily, that I have contemplated the marvels of Creation: my soul has savoured their delights and blessed the Author of their existence." Rafinesque believed that he was the successor of Linnaeus, and he paid him homage, while he complained that he himself had been hindered by neglect and by "feeble rivals," just as Linnaeus had: "Like him I have wrestled against adversity and envy: yet I find in myself my reward and my consolation, the sweet rejoicing attached to the well-considered spectacle of the universe and the study of living things has overwhelmed me with a pleasure unknown to vulgar souls and of which they cannot rob me."

A deluge of words sprang from Rafinesque's fertile and fervid mind. A few years later, when he resided in America, Benjamin Silliman, editor of *The American Journal of Science and Arts,* complained that if he accepted everything Rafinesque sent him there would be no room for anything else. Volume One of the journal contained twelve papers and communications by Rafinesque on a variety of subjects, but by Volume Three there was only one. Rafinesque's papers were also rejected because of repeated denunciations of his science and behavior. The influential Thomas Cooper, professor of Chemistry at Penn and later head of the University of South Carolina, warned Silliman about Rafinesque's unreliability.

With Napoleon relegated to the Isle of Elba, peace broke out and communication with Italian and French professors and naturalists resumed. Whatever judgments were later made of Rafinesque's efforts, at the time of publication the scope and detail of his work and thought were sufficiently impressive for the Academy of Natural Sciences of Naples to send him a Diploma of Honorary Membership, a visible mark of recognition that was important to Rafinesque. This, the first of his honors, whetted his appetite for more, prompting him to nag his peers in other lands about the possibility of becoming a corresponding member of their Academy.

He planned to visit Paris to join his mother and to capitalize on his recently published *Analysis of Nature,* which he had written in French to establish a reputation in France, the center of his intellectual world. Unfortunately, just at that time (1815) Napoleon escaped from Elba, and his bloody, desperate, one-hundred-day adventure unfolded, to end at Waterloo. Rafinesque, "a peaceful man," thought it best to remain in Sicily for a time, but he was determined to leave as soon as he could "where I had so long been detained . . . the disgusting injustice I experienced in Sicily, made me anxious to leave it, I threw again my eyes towards the United States." Precisely which injustices Rafinesque was referring to were left unclear; they were probably of a business nature, for Sicilian functionaries and entrepreneurs may have wanted a share of the profits of a lucrative business established by a foreigner. Without question, he was not treated fairly, neither in Sicily nor in the United States, but ten years had passed and the memory of America was probably becoming rosier as his fortunes in Sicily declined.[17]

Rafinesque rarely wrote of personal matters. Therefore, a rounded picture of Rafinesque is beyond our reach, because description is often based on surmise. Certainly a strong personality, his urge to write and to communicate was powerful, and his celebration of nature and discovery is almost delirious. Yet his autobiography is strangely terse and threadbare, too frequently banal, and rarely illuminating the subject or his world. He does not even give his date of birth, nor is there anything about his marriage, and very little about his family or his personal life. Nothing is known of his home in Palermo or later in Lexington, Kentucky, or his flat in Philadelphia—these were not worth mentioning.

We are surprised to learn about his marriage, and this was only revealed through a will that was later filed in Philadelphia. In 1809, Rafinesque married a Sicilian woman, Josephine Vaccaro (the name Dorothea may have preceded Josephine). It is possible the marriage was not consummated immediately. He considered himself lawfully married while in Sicily "although the decree of the Council of Trent forbade our regular marriage." Perhaps they were not married in a Catholic church, sanctified by religious ceremony. Was he a professed theist or Protestant, or would he not submit to the dictates of the Catholic Church? These questions remain unanswered, though there would seem little doubt that his wife belonged to the Church. When Rafinesque was in America years later, he indicated that his beloved had a beautiful face and blond hair, but her heart was "false," she was "stupid," and he detested her.[18] In 1811 a daughter Emily was born, and in 1814 Josephine gave birth to Charles Linnaeus, who died in infancy. The next year, Constantine left Sicily—a land of "perfidious" women—a description prompted by the fact that when Josephine heard that he had been shipwrecked on American shores and was for a brief time presumed dead, she hastily married Giovanni Pizzalour, a vaudeville comedian. The most casual inquiry would have informed Josephine that her beloved was still alive. Since Constantine was frequently away from home, and his ways were so erratic and unpredictable, it would be no surprise to find that she was glad to see the last of him. Constantine's Sicilian assets quickly disappeared, and two requests that his daughter be sent to him were ignored. Rafinesque saw neither his wife nor his daughter again. After this sad affair, Rafinesque never again formed a close relationship with a woman.[19]

Rafinesque had covered the length and breadth of Sicily, traveling 1,600 miles by carriage and litter, for in Sicily no one except a beggar would travel on foot. However, in America he walked! To the end of his days he was filled with both admiration for and abhorrence of Sicily, that siren land. His *Epitome* of 1814 ends with apologies for the shortcomings of the work and a curse for his tormentors. He had in fact done some remarkable research on fish and plants, and his extensive publications were notable, certainly comparable in quality to most of the work of the time. Frustrated and unhappy, he left Sicily "to vegetate in its willful ignorance." Apologetically, he anticipated his readers' criticism; if they found his work "inconsiderable and incomplete, attribute these defects to the evils of the times and the inaptitude of the Sicilians to understand and appreciate works of such importance, people ignorant of the first element of the sciences can hardly do other than despise refinements and sublimities . . . they cannot suffer a stranger to come and shame them . . . they disparage my discoveries to avoid having to blush at them [for their richness]. I hope to have more justice rendered me elsewhere, the approval of enlightened naturalists will be my sweet reward."[20] To a considerable extent his despair was justified.

Once again Rafinesque, the perpetual, impolitic outsider, seemed to be fleeing for his life while mustering an air of dignified assurance. After spending a decade in exile—his formative years as a natural scientist—away from the major centers of intellectual ferment, it was almost inevitable that this inveterate wanderer would end his days in the New World with its endless mystery and promise. He never saw Europe again.

Chapter 4

AMERICAN SCIENCE AND NATURAL HISTORY IN RAFINESQUE'S TIME

While the young republic was taking shape, American natural historians felt that their first task was to take inventory of the virgin riches of their land, to classify and describe, and to fit the data into a Christian cosmology; for the historians, to describe them was to understand. They became field-workers, exploring the wilderness, bringing back specimens that they preserved and drew, and they established gardens and museums. In this initial period, there was little inclination to experiment when there was so much new to find, describe, and classify in the endless expanses of their country. At the beginning of the century, very few Americans were involved in specialized scientific studies, and almost none were prepared to engage in laboratory investigation. Many decades passed—until after the Civil War—before the first institutions with costly laboratories and great libraries in the European manner emerged.

Education was important, but scientific matters were not pressing. In all aspects of taste and culture America had been docile, as it followed the lead of Great Britain, France, and Germany. With the coming of independence, new attitudes toward government, politics, and science began to emerge, albeit still influenced by Europe. America's most outstanding scientist, Benjamin Franklin, had been dead for a decade, but the country was now miraculously blessed with Thomas Jefferson, a president who was passionately devoted to science and the study of nature, who cherished practical and useful knowledge for the benefit of all citizens. Naturalists and scientists must have been heartened by the active participation and sympathetic regard of someone as important as their president, but even so, Jefferson did not believe that it was the government's function to build universities and support research. President Washington had been so in favor of a national university that he had contributed money for its establishment, but the venture was stillborn because powerful interests believed that in a free society the government should not in any way exert control over education. Without state or federal government support, and with minimal private patronage, there was little chance for science to flourish. In the early days of the republic, the reach of government was kept to a minimum by the exercise of constitutional constraints; "economy in government" was the watchword, with no support even for the building of roads.

Although knowledge of science and of natural history were admired in educated circles, they were not qualifications for high positions in government. Essays on Natural History, a fusion of science and philosophy, were merely something of a literary genre. President Jefferson's talent in science and his wide-ranging curiosity were not appreciated by his political enemies who fulminated:

> Why . . . should a philosopher be made President? Is not the active, anxious, and responsible station of Executive ill-suited to the calm, retired, and exploring tastes of a natural philosopher? Ability to impale a butterfly and contrive turn-about chairs may entitle one to a college professorship, but it no more constitutes a claim to the Presidency than the genius of Cox, the great bridge-builder, or the feats of Ricketts, the famous equestrian. Do not the pages of history teem with evidences of the ignorance and mismanagement of philosophic politicians? . . . But, suppose that the title of philosopher is a good claim to the Presidency, what claim has Thomas Jefferson to the title of Philosopher? Why, forsooth! He has refuted Moses, disproved the story of the Deluge, made a penal code.[1]

Despite the lack of prospects in obtaining a position in government, industry, or in one of the few institutions of higher learning existing in the country, there were those who persisted in following their interests rather than their common sense, choosing the study of nature as their life's work. In 1802 there were only twenty-one faculty positions in science in the country. Candidates of modest means took great risks in committing themselves to a life of science that offered little chance for an adequate living.[2]

To sustain themselves, some naturalists took refuge in a socialist commune (Thomas Say and Charles A. Lesueur in New Harmony, Indiana), while Audubon threw himself into sales and self-promotion. Others joined expeditions, usually to the unexplored West, as scientific specialists. American investigators were mostly gentlemen, earnest amateurs of assured income, whose interest in natural history began when they were boys.[3] The view of the wealthy, retired George Ord, president of the Academy of Natural Sciences of Philadelphia prevailed: "But so fascinating is the study of natural history, so completely does it predominate over other studies that it seems by no means advisable to recommend it to the early attention of youth . . . lest what was intended merely for pastime becomes an occupation."[4] Science, which was not particularly useful, or esteemed by polite society, was tolerable only if one dabbled in it. It took many years for the populace to appreciate the value of seemingly useless knowledge, which ultimately could generate wealth and lead to innovations of direct benefit to society. Between 1815 and 1820, (just when Rafinesque returned to America to stay), the number of professional scientists, scientific institutions, societies, journals, and teaching positions began to grow, giving Rafinesque some reason to believe that he could support himself in the study of natural history.

At the time that Rafinesque arrived in the New World, Americans barely realized that natural history was more than an enthusiastic description of nature, and that for an effective science to evolve, the subject had to be broken down into a series of disciplines—crystallography, mineralogy, geology, botany, and zoology—each with its specialized practitioners trained in rigorous methodology. Most popular throughout the century was botany, a subject that was accessible to all, attracting thousands of "botanizers," amateurs who learned as they collected and traded specimens.[5] Although interest in the wonders of nature was fashionable and widespread, there were no system-building thinkers, no great classifiers, no full-time naturalists attempting to delineate the rational order of the world that underlay nature's grand facade. As useful and supportive as President Jefferson was to the country's scientific effort, he could not be considered a first-rate scientist in the class of the late Franklin. None of the great conceptual advances of the day in science were the fruit of American genius, but rather they were the creations of older, richer, more mature Europeans.

Because few scientists and natural historians were being trained in American schools, the number of graduates capable of becoming professors and investigators was meager. The shortage of accomplished, native-born teachers was so pressing that Thomas Jefferson felt compelled to look to Europe for professors for his newly created University of Virginia (1817), but even so, nationalistic sentiment gave rise to resentment that "qualified" Americans were being passed over.[6] Virtually all the naturalists who were working in America in the first third of the century were not formally trained—Say, Bonaparte, Lesueur, Titian Ramsay Peale, Thomas Nuttall, and Rafinesque. David Rittenhouse, the physicist and astronomer, derived much of his income as a surveyor; it could be said that Jefferson, a part-time scientist, earned his living as president.[7] Not strictly professional, they practiced their science through inclination, and they became qualified through practice.

Here and there, larger institutions were just beginning to support professors who specialized in chemistry, natural history, mineralogy, and mathematics. In the late eighteenth century, natural history and astronomy were taught at Harvard, and universities and colleges began to establish museums, libraries, and facilities for the teaching of science. Benjamin Silliman at Yale lectured on chemistry, geology, and mineralogy. At that time, Philadelphia was the premier center of learning, endowed with the largest libraries, natural history museums, and botanical gardens. In 1816, a Faculty of Natural Sciences was established at the University of Pennsylvania, independent of its medical school, but despite the local wealth, and the high quality of its professors, the school languished and died in 1828 because of a lack of students and a broad base of interest, while the profession-oriented medical school thrived.[8]

For proper training and polish, it was necessary for the serious student of medicine and natural history to spend a few years in Great Britain or on the continent to take advantage of the fine libraries and museums and to attend courses given by brilliant professors. At the same time that European science and scholarship were admired and deemed worthy of emulation, the continent's authoritarian political systems were reviled by American students abroad.

Into the American void in the early nineteenth century came European travelers, cosmopolites, literati, and natural historians like Rafinesque, J.J. Audubon, Frederick Pursh, Thomas Nuttall, Abbé Correa da Serra, Charles A. Lesueur, Charles L. Bonaparte, and Alexander Wilson. Some resided permanently in America, collecting and classifying, while others returned to Europe after many years, taking with them their valuable collections. They were eager to discover an exotic new world, romantics attempting to find a balance between imagination and reason. This interplay, in which intuition played havoc, was most evident in Rafinesque. Men and women came to an unknown America flushed with an enthusiasm that often found lyrical expression in poetry and art. Their world was increasingly dynamic, one in which old, static conceptions were being displaced. Ancient dogma that man was hopelessly limited and must submit to a higher power was questioned. Now the striving of the individual was paramount, and nowhere on earth was individual freedom pursued more vigorously than in the New World.

After humble beginnings, when practical necessity dominated American thought for almost two centuries, scientific effort and achievement began to follow an upward curve. Early in the ascent, science was largely confined to Boston, New York, Philadelphia, Baltimore, and Charleston, but very soon, other centers of learning sprang up throughout the country as the population spread westward and new communities were established. There were always leading men in each new locality, usually from large eastern cities, who were eager to band together to found institutions of higher learning and academies modeled on those they had left behind. Transylvania University in Lexington, Kentucky, where Rafinesque was a professor from 1819 to 1826, was one of these.

The derivative nature of American science and culture was apparent, and yet the notion of American exceptionalism was increasingly asserted; Americans were confident that they were a chosen people in a promised land with a superior form of government. Inevitably they would surpass European achievement in several areas, but at this early time they needed European approbation, and they were overly sensitive to criticism of any kind. The carping Mrs. Trollope[9] and Charles Dickens[10] wrote of the many ignorant, opinionated, and uncouth Americans they had come upon during their travels. They abhorred these ruffians' filthy habits of chewing tobacco, their incessant spitting, their use of the knife as a fork, and horror of horrors, their lack of fine European manners and their obliviousness to the natural superiority of their betters. An overdeveloped sense of individualism and a ferocious entrepreneurial instinct gave rise to embarrassments that filled the newspapers and reinforced the patronizing attitudes of Europeans toward American boors. Samuel Johnson felt that the man who immigrated to America would "immerse himself and his posterity for ages in barbarism."[11]

To Europeans, the boom and bust economy of the United States was emblematic of an immature and unreliable people, but they were quite willing to take their chances on becoming rich through American investment. Those who had lost money or were the victims of American fraud (which included scientists, clergymen, and writers), had no love for the United States. However, America did have its ardent defenders such as Alexis de Tocqueville, and Romantics such as Byron, Chateaubriand, and Rousseau, who extolled the noble savage. Their view of America as exotic was not informed by the brutish experience of life in the wilds or in the barbarities of the new settlements. They would have been appalled by the actual living experience in this land of wonders, where newcomers might spend a winter in a hole in the ground.[12]

The experience of a European traveler is revealed in Francis Sheridan's memoir of a visit to Galveston, Texas, population 5,000, in 1840. He and his traveling companion, General Thomas Jefferson Green, lodged at the Tremount House, Galveston's finest hotel. Six men shared a room, ten by fifteen feet, that contained six "couches," two basins with jugs, and a blazing stove that created insufferable heat. Before retiring, they sat around having a "quiet spit," and during the night the heat, snoring, and other bodily noises were so intolerable that Sheridan preferred to sleep on the freezing balcony. Others found crowded hotel rooms so ridden with bedbugs that they spent the night in the lobby.[13]

Throughout the century, many Europeans, as well as Americans, looked with dismay at a growing, enfranchised rabble—corrupt, ignorant, and ill-mannered—who were having major effects on the outcome of elections and on the temper of American politics. And yet the republican spirit was one of hope, idealistically based on the belief in the perfectibility of man. In the meantime, enlightened liberals on both sides of the Atlantic agreed, as did the Founding Fathers, that only the "best" should lead. Charles Dickens's view of America and its people was jaded, as elaborated on in the travels of Martin Chuzzlewit (and his servant-companion Mark); yet even he could see the potential for greatness. As their ship set sail for England after a nightmarish experience in the new American heartland where they were swindled, Mark asked himself how he would paint the American Eagle. He answered: "I should want to draw it like a Bat, for its short-sightedness; like a Bantam, for its bragging; like a Magpie, for its honesty; like a Peacock, for its vanity; like a Ostrich, for its putting its head in the mud, and thinking nobody sees it. "Breaking in, Martin Chuzzlewit added, "And like a Phoenix, for its power of springing from the ashes of its faults and vices, and soaring up anew into the sky."[14]

Sensitivity to European opinion was not a new phenomenon. The great French savant, Buffon, earned the enmity of Americans when he wrote in 1766 that the plants and animals of the New World, which he believed had come from the Old World, had deteriorated in size and vigor because their diet was poorer, the climate harsher, and the environment was generally inferior.[15] For this opinion, which was, in fact, widely held in European intellectual circles,[16] he was ritually cursed at Fourth of July celebrations. The slur disturbed Jefferson, who went so far as to bring to France and Buffon the skin of a magnificent panther and a stuffed moose, seven feet tall. Inferior indeed! Americans were willing to concede that change could take place in the New World, but not degeneracy, in their land of promise.

In answer to the charge made by the French philosopher and historian, Abbé Raynal, that America has not produced one good poet, mathematician, or a genius in art or science, Jefferson answered that when America would be as old as the Greek, Roman, or European civilizations its productions would be as abundant. Further, considering the great differences in populations of Europe and America, it had done very well indeed, for it had produced a Washington in war, a Franklin in physics, and a Rittenhouse in astronomy, as well as several ingenious inventors.[17]

Americans may have regarded Europe as the ultimate arbiter of excellence and sought European approval of their efforts. But a new voice for a different kind of independence was sounded, especially after the War of 1812, and Americans insisted that they were capable of operating on their own. The 1814 London publication of Frederick Pursh's work on plants that were collected on the Lewis and Clark expedition and were brought to Britain for classification dismayed American naturalists. With the increasing competence and confidence of a growing number of American naturalists, a publishing establishment was created that employed skilled engravers, illustrators, and printers. Expeditions to the frontiers now brought to light large numbers of new species that American naturalists maintained should be described and named by Americans. They had reason to believe that intellectual life could thrive in a democratic republic and that there were effective alternatives to royal patronage.[18]

The public at large became entranced by the wonders of natural history, and the vast array of living forms on the continent became a source of national pride. In Philadelphia, as in other cities, a course in botany offered by the Academy of Sciences was attended by hundreds of students and was repeated to satisfy the demand. Wealthy men like Zaccheus Collins and William Maclure, who supported the work of several American naturalists, provided the Academy of Natural Sciences of Philadelphia with a printing press and an enlarged library. Thus was born the Proceedings of the Academy of Natural Sciences of Philadelphia.[19] From a cadre of pragmatic investigators came extensive surveys of American insects, plants, birds, and molluscs, major publications that were beautifully illustrated and replete with classical allusions and poetry. They were, however, free of speculation and philosophy and without any underlying theoretical rationale.

But the disturbing fact remained that American scientific work was still largely ignored by Europeans. In a letter dated December 7, 1828, William Swainson, the English naturalist and friend of Rafinesque, wrote to an American correspondent that he had been trying to get a copy of the *Proceedings of the Academy of Natural Sciences of Philadelphia* in London but could not. London booksellers had never even heard of the journal—one of the very few major American scientific publications of the time.[20]

Chapter 5

RETURN TO AMERICA 1815–1818

In July 1815, Rafinesque sailed from Palermo on the *Union of Malta* destined for New York. He carried with him "a large parcel of drugs and merchandize, besides 50 boxes containing my herbal [20,000 specimens in 2,000 species], cabinet collections, and part of my library. I took all my manuscripts with me, including 2,000 maps and drawings, 300 copperplates, &c. My collection of shells was so large as to include 600,000 specimens large and small. My herbal was so large that I left part of it."[1]

The trip was painfully long, lasting over three months, with a three-day stop at Gibraltar that allowed Rafinesque time to examine the local flora. When the sea was calm, Rafinesque was able to study aquatic life, but the ship soon ran into a fearful storm that cost its masts, and a brig within their sight slid under the waves. Rafinesque believed he had seen his last day on earth. Battered, the ship limped into port on the island of St. Michael in the Azores, where solicitous British and American consuls treated them kindly. While the ship was being refitted, Rafinesque was able to inspect the flora and the volcanoes of the island. He was delighted by the Azores and would gladly have stayed a month, but the ship was quickly put into shape, and soon they were off for America. Again, they were dogged by a violent sea that forced the crew to throw their cannons overboard, and adverse currents slowed their progress to a point where the last of their supplies were consumed as they approached the American coast. Then the real trouble began.

They sighted Cape Montauk as westerly winds were blowing, making it feasible for them to put in at Newport, Rhode Island, for food and water. But as they headed for safe haven the wind shifted. A strong northeasterly wind now drove them southwest into Long Island Sound toward New York. That night, engulfed in fog, they ran onto underwater rocks between Fisher Island and Long Island:

> Our ship filled fast and settled down on one side; but without sinking, being made buoyant by the air of the hold. We had merely the time to escape in our boats, with some difficulty. . . . Having left the wreck we rowed towards the lighthouse of New London then in sight, and reached it at midnight: thus landing in America for a second time, but in a deplorable situation. I had lost everything, my fortune, my share of the cargo, my collections and labors for 20 years past, my books, my manuscripts, my drawings, even my clothes . . . all that I possessed, except some scattered funds, and the Insurance ordered in England for one-third of the value of my goods. For some days after I was in utter despair.[2]

Hope of salvage was soon cut short when the mastless ship sank while being towed, taking down with it all of Rafinesque's possessions, including medicinal plants he wanted to sell to raise cash. Oddly, soon after the time of the disaster, Rafinesque wrote to a friend in Tuscany that he swam away from the wreck and while doing so observed several new genera and species of fish and aquatic plants. At this time of extreme anxiety, he seems to have been delusional. Three years later, in a letter, he related a similar version of the story, but not quite as bizarre. Describing a species of cod, he wrote: "I observed several of these fish . . . on board of a fishing-smack off Point Judith, Rhode Island, the very same day of my unfortunate shipwreck."[3]

The hostility Rafinesque had engendered on his first visit to America was still alive, even after ten years. Easygoing American collegiality had not been Rafinesque's style, and for this he paid a price. Despite his European formal manners, he could barely conceal his erratic un-American enthusiasms or his likes and dislikes. Now, at this terrible time, a pathetic but defiant cry arose from the depths of his being: "Some hearts of stone have since dared to doubt of these facts or rejoice at my losses! Yes, I have found men, vile enough to laugh without shame at my misfortune, instead of condoling with me! But I have met also with friends who have deplored my loss, and helped me in need!"[4] Rafinesque was impoverished by this catastrophe to the end of his days. Poor Rafinesque could hardly cross the Atlantic Ocean without suffering a shipwreck or other disaster.

The country Rafinesque returned to in 1815 was different from the one he had left ten years previously. At the very dawn of its industrial revolution, the nation was richer and more sure of itself. Botany in the United States had advanced remarkably during Rafinesque's decade in Sicily, where even from afar he kept publishing descriptions of American species in *The New York Medical Repository*. There were many able botanists such as Frederick Pursh, Thomas Nuttall, Stephen Elliot, Jacob Bigelow, and Henry Mühlenberg, who were making significant contributions, and along with the Michaux, father and son, had each published a useful, comprehensive botanical catalogue. These *Flora,* describing plants from various parts of the United States, facilitated their identification by amateurs and professionals and popularized the study of plants in the great outdoors. Plants, usually listed according to the Linnaean system, were described in sufficient detail to serve as a key for the identification of the genera and species of new forms, almost always a difficult task. Essential information about plants was also provided, such as habitat, characteristics of growth, time of flowering, size and descriptions of leaves and flowers, including color and number of petals and stamens. These works provided a solid foundation upon which future North American botanical studies could develop.[5]

Rafinesque found his way to New York where, with the help of Samuel L. Mitchill, a scientific leader there, he made new friends—"learned" men—including Governor DeWitt Clinton; Dr. David Hosack, Professor of Botany and Medicine at Columbia College and attending physician at the Burr-Hamilton duel, whose garden stood on the present site of Rockefeller Center; Caspar Wistar Eddy, a botanist; and others. Mitchill was not only a physician who had married well, but he was also a man of broad learning who served in the U.S. House of Representatives and in the Senate. For a time, he proved to be a guardian and steadying influence on Rafinesque. A word from Mitchill was sufficient for Rafinesque to become the tutor to the wife and three daughters of New York patrician Robert L. Livingston. Much of the snowy winter was spent at the Livingston family's Clermont estate on the banks of the upper Hudson River, where Rafinesque gave instruction in Italian, botany, and drawing, a task that left him sufficient time to read, think, and write in their fine library. This idyll suddenly ended when Mrs. Livingston became ill and the family moved to the more favorable climate of South Carolina. Rafinesque, suddenly dispossessed and left to care for himself, decided to visit his colleagues in Philadelphia.

Fortunately, a Philadelphia Quaker merchant, the patrician Zaccheus Collins, came to support the luckless Rafinesque. Collins, sharing Rafinesque's abiding interest in botany and natural history, proved to be a faithful friend, advisor, and patron. Sensitive to Rafinesque's nature, Collins encouraged him to travel and collect and supported him in his endeavors. Rafinesque spent the spring of 1816 exploring New Jersey with Mitchill and Alden Partridge, picking up plants, shells, and fish. Later, he looked closely at the flora and fauna, minerals, and fossils of the upper Hudson River, drawing and mapping "nearly its whole course." He also explored northern New York state and Vermont, after which he returned to New York by steamboat. For Rafinesque, this activity, however strenuous, was therapeutic.

With Manhattan and Brooklyn as his headquarters, he gradually recovered from his great loss. He received some insurance money, started a commercial venture shipping material to Sicily, and had every intention of settling in New York "in trade," but once again, he was frustrated by a series of commercial calamities. He was cheated out of his remaining assets in Sicily by a "perfid Sicilian" whom he had considered a friend; there was also "the bankruptcy of a house in New York, law suits and other troubles."

One year after Rafinesque's disastrous arrival in America, he published a remarkable document, *Circular Address on Botany and Zoology,* in which he wrote of the shipwreck that left him in a "destitute state." "This dreadful misfortune has not, however, impaired my zeal; I am determined to begin again my labours." What followed was a worldwide appeal to all investigators to exchange specimens with him—plants and animals that he would identify and name—and if at all possible he would provide them with any specimens they wanted. He appealed to booksellers to send him books, and in return he would supply them with his writings so that they might sell them, with a commission of 10 percent going to Rafinesque. He importuned: "Whatever be your situation in life, and wherever is your abode, I hope we may be of use to each other. . . . Ask me in return anything in my power to bestow, plants, animals, books, my works &c." Rafinesque, though penniless, attempted to create a vast scientific community—a global brotherhood.[6]

Given his resources, this rational plan was hopelessly impractical, but there was a religious fervor in his pronouncements: "I unite to the most glowing ardour for the knowledge of nature, the most ardent desire to promote its study, by all the means in my power." His ambitious yet touching entreaty ended on an idealistic note, a libretto that would not be out of place in *The Magic Flute*: "If we are already united by a mutual love of nature, and pure zeal for the investigation of the wide field of natural sciences, let us strengthen the ties of our union." Only a few responded favorably to his call, yet even so, Rafinesque was always the center of an extensive, lively correspondence.[7]

Rafinesque issued a prospectus of *Annals of Nature and Somiology of North America,* an ambitious publication that would describe all known plants and animals, under the headings of Nomenclature, Diagnosis, Description, History, Properties, and Peculiarities. *Annals* was to be a serial publication of five thousand numbers over eight to ten years. Conveniently, it would be sold in sections so that those with special interests—"farmers, gardeners, ladies, sportsmen and anglers"—need only buy parts of the whole. Every detail of the production of each volume and the schedule of payment and delivery were described. Rafinesque, an admirer of French science, a latter-day *encyclopedist,* the ultimate optimist, felt that if he could set down a rational plan on paper there was no obstacle to its consummation. Sober men would hardly take seriously his absurdly ambitious schemes, no matter how meticulously described. For instance, he wrote an astonishing account of the date tree or palm, its growth characteristics, cultivation, and fourteen uses of almost every part of the tree.[8] Despite the impracticality, he never missed the opportunity of emphasizing the uses of nature.

He sought refuge in the world of the intellect by joining the local Literary and Philosophical Society and was a very active, founding member of the Lyceum of Natural History of New York, the precursor of The New York Academy of Sciences. With Mitchill as president and the Academy of Natural Sciences of Philadelphia as a model, Rafinesque was deeply involved in the first efforts of the Lyceum to establish a library and a museum and to initiate a series of lectures. He was a member of numerous committees, including the lecture committee, and eagerly delivered the very first scientific report at a meeting of the Lyceum.[9] Through Rafinesque's efforts, the *Annals of Nature,* a semiofficial journal of the Lyceum, was founded, to be published four times each year for four years. Each number was to contain sixty to one hundred articles on the natural history of the United States, but the *Annals* died after only one number, probably because members could see that the entire journal would be taken up by Rafinesque, even though William Swainson praised the publication.[10]

Charitable critics considered Rafinesque to be bizarre and tactless, but all considered him remarkable in the breadth and depth of his knowledge and experience, and his industriousness was beyond belief. Kindly men like Amos Eaton and John Torrey, both New York botanists, were well-disposed toward him, but their tolerance had limits. Torrey wrote to Eaton, "He is the best *naturalist* I am acquainted with, but he is too fond of novelty. He finds too many things. All is new! New!"[11] Whatever the exasperation expressed, Torrey corresponded extensively and in detail with Rafinesque over many years. Eaton wrote to Torrey:

> I am glad Mr. Rafinesque has not set you all wild. Why can not he give up that foolish European foolery, which leads him to treat Americans like half-taught school boys? He may be assured, he will never succeed in this way. . . . His new names with which he is overwhelming the science will meet with universal

> contempt. Cannot some friend induce him to return to sober reason, and thus make himself highly useful and much esteemed?[12]

> I have defended him in New England, until I am ashamed to mention his name. His name is absolutely becoming a substitute for egotism. Even the ladies here, often adorn their witticisms with the name of Rafinesque, applied in the same. They talk of the science of Rafinesquism; meaning the most foolsome and disgusting manner of speaking in one's own praise.[13]

And these were his friends!

There were disturbing questions about his scientific judgment and his veracity. In 1817 he published a botanical work *Florula Ludoviciana* [*Flora of Louisiana*], dedicated to the governor of New York, DeWitt Clinton, a reworking and loose translation of the botanical component of a book on travel by C.C. Robin, a Frenchman who had explored the Gulf Coast from Louisiana to Florida. Though not a botanist, Robin carefully described many of the plants he had come upon. Looking over Robin's account, Rafinesque found "many blunders in nomenclature and classification," but "several accurate descriptions and valuable additions to the knowledge of plants."[14] From his armchair he took it upon himself to identify Robin's plants and to classify them, and in doing so he came up with thirty new genera and 196 species. Criticized for classifying plants he had not even examined, he claimed that while in France he had seen Robin's collection of Louisiana plants. But in fact, Robin never formed a collection of dried plants of any size; Rafinesque was caught in a lie.

Rafinesque's contemporaries abhorred the *Florula Ludoviciania,* and some have claimed that this exercise marked the true beginning of Rafinesque's decline. In his 1825 address on the state of science in the United States, James DeKay, president of the New York Lyceum of Natural History, did not mention Rafinesque by name but stated that the *Florula Ludoviciana* was "a mere compilation from the loose and inaccurate notices of the Abbé Robin."[15] Torrey wrote to Eaton about the Louisiana work:

> This work is the most curious medley I ever saw. The author without ever being in the country whose plants he describes, has discovered 50 or 60 new species. . . . I expect he will soon issue proposals for publishing the botany of the *moon* with figures of all the new species.[16]

> Raf. certainly deserves to be ridiculed—his vanity is absolutely intolerable. . . . Is it not preposterous for a man to pretend to write a flora of a country he

never even visited. If Raf was to travel in Louisiana himself & the very same ground that *Robin* did he could not find the plants he has described.[17]

In Rafinesque's obituary written by Asa Gray,[18] his seemingly dispassionate but derogatory comments about the *Flora of Louisiana* and other works brought together the criticism that had been made over the years. The critique was widely circulated, unquestioned, and enormously influential, for Gray, a rising botanist, was regarded as the standard keeper of the field—with some good reason. Later, however, botanists were more forgiving, even by the strict standards of descriptions of genera and species that were subsequently adopted. Rafinesque's descriptions are adequate and valid, and the plants he listed are for the most part recognizable today. He had attempted what Linnaeus and others had done without criticism—classification on the basis of the published descriptions of others.[19] However, Gray's damage was telling, even after many years.

Francis W. Pennell, a botanist free of the prejudice of Rafinesque's contemporaries, commented on the *Flora of Louisiana,* 125 years later and judged that the names in this work were "validly presented" by Rafinesque according to "strict rules as to what counts adequate publication of new species and genera," and that almost all the plants can be identified from the descriptions of Rafinesque and Robin. Though critical of Rafinesque's all too brief and "unconventional" descriptions, and of his habit of finding everything "new," he was "amazed" at his "versatility" in working with "sponges, insects, crustacea, fishes, serpents and mammals."[20] In 1957, another botanist wrote in the botanical journal *Rhodora*: "I am confident that most of the species of the *Florula Ludoviciana* can be satisfactorily identified. Rafinesque's descriptions are more adequate than Walter's in *Flora Caroliniana* . . . which has been spared the obloquy awarded to Rafinesque."[21] In other words, Rafinesque was a special target. Pennell also reminded us, as does Cain,[22] that the state of the art at the time was rather primitive, binding rules had not been established, and the requirement for type specimens lay in the future.

In this 1817 *Florula,* Rafinesque identified a vine and created a genus for it, *Bradburya,* after his good friend John Bradbury, a plant collector who supplied him with many specimens. A year later the same plant was given the generic name *Wisteria* by Thomas Nuttall—without explanation—and this name has been conserved. The famous vine was named after the influential Caspar Wistar, Professor of Anatomy at the University of Pennsylvania and Thomas Jefferson's fellow paleontologist. The establishment had blatantly disregarded Rafinesque's claim to priority without explanation.

Rafinesque objected to naming plants after people who were not botanists, although he himself did so, upon occasion. Sometimes Rafinesque's renaming of species worked; looking over the description of prairie dogs in Meriwether Lewis's accounts (just as he had with Robin's), Rafinesque assigned them the name *Cynomys* (dog-mouse), which is still accepted. He seems to have scrutinized Lewis' descriptions with care, for from these he also discerned six new species of Oregonian fir, including Douglas fir, none of which he ever saw.[23]

In his time, many regarded Rafinesque as a pariah, and even a fatherly, sympathetic botanist such as Amos Eaton warned a friend that to use Rafinesque's binomials would ruin his work.[24] Eaton was occasionally fed up with Rafinesque, but along with John Torrey remained one of his few defenders. At times, Rafinesque appealed to Torrey to mollify a victim of his barbed criticisms. In 1835 Eaton wrote: "Even those who are disposed to pronounce Mr. R. an extravagant enthusiast, all agree, that he is a scholar of the first order; of vast reading and great classical learning. His nice discriminating talents have never been questioned. . . . '*Shades of variety' as evidence of specific differences* comprise all his supposed scientific heresies."[25]

Most damaging to Rafinesque was the fact that almost none of his names were listed in any botanical indices, including the comprehensive *Index Kewensis,* a master list of the plants of the world, nor has the International Botanical Congress approved them in relatively modern times. Indeed, even in recent years, an effort by an impassioned critic was mounted to officially remove Rafinesque's work from the botanical record—but saner heads prevailed. Without official listing, his names were on the road to oblivion. Toward the middle of the nineteenth century, the pendulum swung from exclusion to a qualified rehabilitation of Rafinesque's reputation. By the end of the century, though botanists were fully aware of his shortcomings, several botanists and zoologists praised him and thought he was unfairly maligned. They had, of course, had no personal contact with the man.

As Rafinesque's reputation was rehabilitated toward the end of the century and his papers were critically reevaluated, it was apparent that a considerable amount of his work did, in fact, have validity and had been "overlooked." His binomials, hitherto unknown, kept popping up. Pennell listed eighty-four, and Merrill accounted for a total of 740 overlooked names.[26] It was incumbent according to the rule of priority that the status of accepted binomials, even longstanding ones, should be reconsidered in the light of Rafinesque's work, and if Rafinesque's names were found valid, they should displace those assigned years after and in common usage; the longstanding names would be officially reduced to *synonym* status as recorded in the *Index Kewensis*. In some instances this occurred, but the process has been dogged by uncertainties. It was only in 1929 that some entries from Rafinesque's 1840 *Autikon Botanikon* entered the *Index Kewensis.*

Merrill has examined nearly all of Rafinesque's botanical papers and has estimated that there may be as many as 1,500 "overlooked and unrecorded generic and specific names." After this heroic task, he suggested that an intensive effort to establish a complete list would be a worthy project, for there was much that was good and salvageable.[27] In this spirit, Stuckey, in 1971, investigated the 275 specimens of vascular plants remaining at the Academy of Natural Sciences of Philadelphia belonging to Rafinesque.[28] Stuckey estimated that overall Rafinesque had proposed "about 2700 generic names and over 6700 binomials in approximately 1000 known publications," but careful, objective evaluation (when possible) has resulted in the acceptance of only thirty generic names and only 150 binomials—a poor record indeed.

The "nomenclatural chaos" spawned by Rafinesque's work, sometimes reliable and at other times not, (often botanists could not decide), led some people to declare that his names be assigned homonym status or that his work should be entirely and officially discarded, for it was not worth attempting to sift the wheat grains from so much Rafinesquian chaff. Others, especially those whose pet names would be displaced, thus depriving them of a measure of immortality, had their good reasons for wanting his work ignored. Even Merrill, sympathetic though he was, concluded that Rafinesque had done more harm than good, an opinion shared by investigators in other fields to which Rafinesque contributed over the next two decades—ichthyology, conchology, and ethnology. Because of the rule of priority and despite the fact that so many of his names were rubbished, there remain at the present time a very large number of plants and animals with *Rafinesque* in their names. Taxonomists notice them for their euphony and for an echo of the meadows and forests of an early America.

Though Asa Gray could be ungenerous to those whom he considered to be without credentials or in any way threatening, he did temper his criticism of Rafinesque with something that approached tenderness. Almost grudgingly, he seemed to recognize Rafinesque's uniqueness and genius and his gift of imagination. Rafinesque's notion of perpetual change in living forms may have been dismissed as foolishness, but it also gave Gray pause: "It is indeed a subject of regret, that the courtesy which prevails among the botanists of the present day, (who are careful to adopt the names proposed by those who even suggest a genus,) was not more usual with us some 20 years ago. Many of Rafinesque's names should have been adopted; some as a matter of courtesy, and others in accordance with strict rule. But it must be remembered, that the rule of priority in publication was not then universally recognized among botanists"[29]

At a time when support for science was meager, those without a private income deemed a university or college appointment a highly desirable prize. Teaching was their prime responsibility, but there was always time to indulge one's interests in nature. With the death of Benjamin Smith Barton in 1815, two professorships in the College and Medical School of the University of Pennsylvania became available; Rafinesque eagerly applied for the professorship of botany and natural history. From Clermont, the grand Livingston estate on the Hudson River, he wrote asking for the position so that he might teach the natural method of classification of plants and become a second Linnaeus. In his letter, a blend of pomposity and self-debasement tinged with desperation, he pressed his case hard. He appealed to the influential Dr. John Mease to help him obtain this position, adding that if he didn't get the job, "I shall feel myself as doomed to neglect as Linnaeus was for a while."[30] Samuel L. Mitchill of New York had written a letter of recommendation praising Rafinesque's "talent and assiduity . . . high requirements, singular industry and indefatigable zeal, . . . his good manners and conciliatory temper . . . a valuable acquisition to out country."[31] Despite his qualifications, the array of his publications, the numerous societies to which he belonged, and the vigorous mustering of support from influential people, to no one's surprise the job went to Barton's nephew and student, W.P.C. Barton.[32] Nothing could overcome the power of the influential Philadelphia family in matters such as this. In the next few years Rafinesque attempted without success to join state-sponsored geological surveys, and an appointment at a New York university also failed to materialize. Regardless of written and official recommendations, what people were saying about him no doubt ended any chance he had of obtaining suitable employment.

When Thomas Jefferson's second term as president of the United States came to an end, he returned to Virginia to restore his property after years of neglect. The idea of a University of Virginia grew in his mind, and by 1819 it became a reality. Hearing that professors were needed, Rafinesque wrote both Jefferson and the Board of Trustees, eagerly offering his services and reminding Jefferson of their meeting in 1804.[33] He offered to teach natural sciences, French and Italian, materia medica, natural philosophy, mineralogy, geology, physics, geometry, map drawing, and political economy. He was willing to set aside part of his salary to purchase books for the university library and offered to donate his personal herbarium and library. Again, his references were impressive—De Witt Clinton, Zaccheus Collins, and Samuel L. Mitchill. A series of evasive letters containing stock excuses from Jefferson (delays in building, lack of funds, decision lays in other hands, etc.) and importuning letters from Rafinesque were exchanged until 1824, but in the end Rafinesque was not hired. Again, most probably, confidential criticism of Rafinesque—some serious, and some mean-spirited and malicious—had reached Jefferson's ears. However learned and brilliant Rafinesque was, he was still too much of a risk, an unpredictable burden. Jefferson was polite but firm in thwarting the applicant.

An important aim of the Lyceum of New York was to collect, identify, and characterize the entire flora and fauna of the New York area, and no one was more eager to fulfill this task than Rafinesque, who continued his explorations of the Hudson River (seeking its source), the Catskills, and Long Island. His time was taken up with relentless travel and "herborization," but he managed to write a remarkable number of papers. From 1817 to 1819, Rafinesque published more than forty articles in *The American Monthly Magazine and Critical Review,* a semiofficial journal of the Lyceum of New York. As one of the editors of this journal, he was in a position to publish endless lists of new genera and species of plants and animals—water snakes, prairie dogs, crustacea and fish.[34] A review of the progress of American science since the turn of the century was masterful—detailed, informative, and well-written in the characteristic embellished language of the time. Writing as an American with an international outlook, he reviewed the achievements of individual naturalists, institutions, museums, libraries, gardens, and books—all American.[35]

Rafinesque identified himself completely with America, and there was no one there quite like him who knew as much as he did. He wanted American scientists to excel. And he felt that in order for them to be as good and as imaginative as Europeans, they would have to discard the outdated and the unworkable and welcome the new. Earnest criticism offered, however painful, should provide an opportunity to improve. Criticizing *A Manual of Botany for the Northern States,* by his good friend Amos Eaton, Rafinesque wrote: "as his whole Zoological book proves . . . he is forty years backwards in the science of Zoology, as he is 30 years backwards in Botany, and about 20 in Geology. But this is not peculiar to him, it is the fate of one half of our Naturalists, Botanists and Geologists. The daily increase of knowledge and improvement in science is despised or neglected by them as useless innovation! While all the world, and all the sciences move forward, they would keep those they teach or cultivate at a stand! it is all in vain, and time will show it."[36]

He found Eaton's book to be "a mere compilation. . . . This unnatural, incorrect, difficult, puzzling, indelicate and obsolete [sexual] system prevails yet in the U.S." He criticized the language, called attention to many errors, and concluded that the work was "practical and useful but by no means classical." His informative review of Eaton's *Index to the Geology of the Northern States* was equally dismissive—a blend of praise and serious criticism.

William Maclure's masterwork *Observations on the Geology of the United States of America,*[37] was treated kindly but he could not resist noting at one point the "preposterous conclusions of Mr. Maclure." Rafinesque must have known that the wealthy geologist and educator William Maclure, president of the Academy of Natural Sciences of Philadelphia and its major benefactor, was a generous promoter of American science and the New Harmony Colony in Indiana. Surely in the real world he deserved a little flattery.

In a review of W.P.C. Barton's *Flora Philadelphica Prodromus*, Rafinesque pointed out that the author had missed many known plants and that he should consult with the reviewer for a complete list. After berating Barton for holding to the discredited Sexual (Linnaean) System of classification, he dismissed the work. "Let us hope that our botanists will avoid the faults it has been needful to point out."[38]

He was harshly critical of the publications of Thomas Say on the insects of North America,[39] and Frederick Pursh's landmark work on plants of North America.[40] He castigated the latter author for his omissions and for not giving sufficient credit to others like himself and Caspar Wistar Eddy of New York. Rafinesque wrote: "Ignorance stamps a degree of imperfection on the whole work. . . . The errors, blunders and misnomers scattered throughout the whole work are numberless . . . to change a good name into a bad one is the most absurd temerity. Yet such absurdity has claimed the preference of Mr. Pursh: we would advise him therefore, as well as those who may be inclined to follow his authority, to go to school, and begin to spell botany, as school boys do their letters." He then went on to list *by number*, forty-three glaring errors, each followed by a correction, and he catalogued many of his own papers describing species that Pursh had overlooked or ignored. In truth, some of these references were in obscure journals, difficult or impossible to consult. Rafinesque knew the literature extremely well (when he wanted to), and he insisted that others should do the same to eliminate errors in assigning priority. In a memoir of Pursh's life and work, Ewan states that many of Rafinesque's criticisms were valid.[41]

Rafinesque called *Introduction to the Ichthyology of the United States* by his patron Samuel L. Mitchill—upon whom he was so dependent—"interesting and authoritative . . . a classic labour . . . although defective in many respects, by a want of synonymy, ignorance of new genera, wrong reunion of species and imperfect descriptions of many."[42]

Careful to not overlook his friends in Philadelphia, he reviewed an issue of the *Journal of the Academy of Natural Sciences of Philadelphia*, and after making negative remarks about how so few Philadelphians actually contributed to the work, he concluded imperiously, "It will be perceived that implicit confidence is not always to be given to the labours of the Academy; but we trust that the published facts and descriptions are correct and to be depended upon."[43] A powerful Philadelphian, George Ord was enraged by Rafinesque's impatient castigation: "Your *Ovis montana* is not a Sheep. I told you so when you exhibited a horn at the Academy."[44]

Rafinesque reviewed Thomas Nuttall's *Genera of North American Plants* in *The American Monthly Magazine and Critical Review* favorably, praising its "neatness of execution, its appropriate plan, convenient shape and cheap price . . . the whole includes a more correct account of our genera than has ever been published." He had many good things to say about this original, "superior" work that he believed to be far more than a mere "compilation" of names. But then he went on to berate the author for using the faulty, outmoded sexual system of plant classification as a disservice to botany, and listed 142 mistakes, which he corrected, as if he were helping a bright, but misguided student. The names of some plants were identified by the reviewer as "wrong," "bad," "absurd," or "inadmissible," and proper names were proffered. He took the liberty of reclassifying some specimens and adding some of his own plants to the list that had been overlooked by Nuttall.[45] Rafinesque was astounded that Nuttall took offence. In a letter to John Torrey in New York, Rafinesque wrote, "[T]ell him [Nuttall] I esteem him highly as an accurate Botanist partial to Natural affinities, he ought not to be offended at my review of his work, it was all for the good of science and justice sake."[46] Rafinesque was more than willing to repair any wounded feelings he had caused and to forgive and forget the insults of others. In 1832 Rafinesque named a new species of sedum *Sedum nuttallianum* (Nuttall's Stone Crop). After Rafinesque's death, Nuttall graciously returned the sentiment by naming a genus of the *Compositae* family after him.

Knowing American scientists well, Rafinesque believed they had the ability to succeed, but at this stage in history they were deficient. In his review of American science of the period 1800 to 1817, he admitted that Americans had not made as many discoveries as the English, French, or Germans but had "increased somewhat the general store of science," and had "added to the physical and natural knowledge of our country." Implicitly, this mild praise, defensive in nature, was a goad to improve provincial American science by raising its standards to a level beyond criticism. Rafinesque made judgments, like a stern schoolmaster who had the right, indeed the responsibility, to scold and to point out errors—relentlessly. Yet Rafinesque usually ended his reviews with words of encouragement and even praise, as a good teacher should. He would suggest ways of improving the work, and even offered to help. How could Americans not be grateful when told that their science would flower only after imperfection was recognized and eliminated?

The review was Rafinesque's dangerous weapon, for there was always the threat of condemnation if an author did not cite Rafinesque. Though not viciously critical *ad hominem*, in actuality, he often cursed those who offended him by ascribing animal characteristics to them—"croaking frogs . . . snail of science"—more comical than mean. He was a missionary, a gadfly, who ultimately expended himself living under the massive and overwhelming burden imposed by his opinions, his behavior, and his credo: "Every single new species or new genus discovered or introduced is a conquest made by knowledge over nullity."[47]

American scientists were happy to ignore or denounce the impolitic, hapless Rafinesque, a man who was not quite one of them. On the other hand, Louis Agassiz, another European immigrant steeped in the scientific culture of Europe, was vastly more successful a few decades later in setting standards for American science and moving it forward. Not only was he immensely able and articulate, he was also a masterful politician who knew how to befriend the right people. Agassiz, who defended Rafinesque, never had personal contact with him, for he came to America six years after Rafinesque's death.[48] The remarkable disparity in the reception of these two men lay in the stunning differences in their diplomatic and administrative skills and in their social graces.

In keeping with his background in both commerce and botany, Rafinesque was remarkably authoritative about the economic aspects of agriculture. He wrote about the tea bush because there seemed to be a chronic shortage of tea in America, which resulted in $12 million leaving the country each year, most of it in much-needed silver coin. He urged Americans to learn about how to import various kinds of tea plants from China and provided information about their cultivation in North America. He accepted the fact that the drinking of tea was a "bad habit" and "useless," and he referred to tea as "that pernicious leaf," but since the habit could not be eradicated, it was worth trying to remove its economic sting by cultivating five species of tea plants in the United States, as he claimed they had been in France.[49]

Rafinesque also regarded "stinking" tobacco as an abomination but was forced to accept its widespread use. His words anticipated those of the modern antismoking zealot: "We have borrowed from the Indians the filthy and vicious custom of smoking or inhaling the vapor of a pernicious weed, a narcotic poison. . . . Whoever smokes pure tobacco habitually, is a selfish vicious man, particularly if he throws the stinking smoke into the lungs of whoever chances to be near him." One almost expects warnings about lung cancer and heart disease, but there are none because his objections were mainly aesthetic, and his solution was the use of a fragrant tobacco or a partial substitution with the leaves of certain plants and trees.[50]

He wrote about the practical details of the cultivation of corn, wheat, hemp, and tobacco, and the marketing forces of the time—the cost of cultivation, expected sale price, and profit per acre.[51] He discussed the conversion of woody substances to sugar and gum,[52] the processing of pumpkin seeds and vegetables for their oil, and the spraying of trees with sulfur to kill insects.[53] Harking back to his family business in a paper on commerce, Rafinesque outlined how one organizes a trading voyage—fitting out a ship, hiring a crew, analyzing the demand for a product, and choosing a cargo to maximize profit.[54]

By 1818 there was little to hold Rafinesque in New York. At thirty-five, his career was not flourishing, for he held no paying position and he had no reliable means of support. He was surrounded by a growing body of the disaffected, who could make his life wretched, and he was denied access to the most important botanical collections, essential for the working taxonomist. The obvious solution to the problem was to leave it all, but he could not refrain from burying himself in his work, maintaining a great interest in the Lyceum and keeping in contact with a few friends, who provided him with enough money to satisfy his frugal requirements. Seeking what he considered a friendlier environment, he left New York for Philadelphia, and from there he began a grand tour of the West with all of its unknown and unclassified abundance.[55] John Torrey wrote to Amos Eaton: "Rafinesque has just started on a three-month expedition. . . . You can imagine how many new discoveries he will make. He was almost Crazy with anticipation before he left here."[56]

Chapter 6

GOING WEST
1818

In May 1818, Rafinesque began a two thousand mile westward journey from Philadelphia. A stage took him to Lancaster, and from there he walked, crossed several ranges of the Alleghenies on his way to Pittsburgh, collected rare plants and shells, studied the geology of central Pennsylvania, and inspected coal mines. Throughout his career, he passed back and forth between the lonely peace of the countryside and the tensions of the city, and though he considered himself an urban sophisticate intrigued by commerce, the locomotive, and steam power, he reveled in a pastoral setting.

Pioneers, trekking through the wilderness to western lands, required maps of the main thoroughfares such as the Ohio River to reach their promised land, and fortunately the multitalented Rafinesque could fulfill that need and earn a few dollars, for he had become proficient at surveying and mapmaking in Sicily and along the Hudson River. He contracted with a Pittsburgh publisher to survey and make a map of the Ohio River from Pittsburgh to the Wabash River, describing the physical features of the valley, its islands and rapids, and the towns along its shores.

In Pittsburgh, Rafinesque and a group of congenial Frenchmen bought a flat, covered boat to take them down the Ohio River at a leisurely pace. Oddly, most of the important naturalists who explored the land beyond the Alleghenies in the early nineteenth century were French or Americans born in France—Bonaparte, Lesueur, Audubon, and the Michaux (father and son). To this list can be added Rafinesque, who surveyed and botanized as the boat drifted down the Ohio, studying and drawing many kinds of fish. This pioneering work was published in his classic *Ichthyologia Ohiensis* of 1820. In all, he described about one hundred new fish, bringing the total number of known North American species to five hundred. Rafinesque's interest extended to plant and animal fossils, which he recognized as markers for identifying and dating the geological strata in which they were found.

Having completed the first part of his voyage, when the boat arrived at Cincinnati Rafinesque walked one hundred miles to Louisville, Kentucky. He spent about two weeks studying the flora and fauna of the region, and it was at this time and place—the Falls of the Ohio River at Shippingsport (now within the limits of Louisville)—that Rafinesque felt himself to be in an earthly paradise abundant in flora and fauna that was his to describe for the first time. He specified 583 three new animals, including "8 new Species of Bats and 10 New Species of Rats or Mice"; he also described 125 new plants.[1]

Rafinesque then took passage on a very slow keelboat that stopped at almost every town along the river, permitting him to collect specimens until his pockets bulged, and he was weighed down by a large bundle of plants on his back—an odd sight indeed! Eventually he reached Henderson, Kentucky, the home of John James Audubon, who at the time was earning a living operating a mill and a general store. Rafinesque, restless and impatient, had had enough of plodding boat travel and was only too happy to abandon it. He was eager to meet Audubon, whose illustrations of wildlife were of particular interest to him, because the birds he painted rested on shrubs and plants that he could identify—perhaps there were some new species among them. The meeting of these men in the late summer of 1818 was recounted by Audubon in his *Ornithological Biography,* and the story has been frequently told as a light and humorous backwoods tale.[2] The encounter can be looked upon quite differently—a dark tale that reveals much about Rafinesque, but more than Audubon would like known about himself. The two met for the first time when Rafinesque, a ragged "odd-looking fellow," asked Audubon, whom he had never met, where Audubon could be found. Upon learning that he had indeed found his man, Rafinesque presented him with a letter of introduction— "My dear Audubon, I send you an odd fish which you may prove to be undescribed." Audubon thought the man was delivering a fish. When he asked where the odd fish was, Rafinesque said, "I am that odd fish, I presume." Audubon felt "confounded and blushed," perhaps because he should have recognized whom this well-known stranger was.

Audubon, struck by Rafinesque's bizarre appearance, as were the boatmen, described the scene: "He pulled off his shoes, drew his stockings so as to cover the holes about his heels telling us all the while in the gayest imaginable mood that he had walked a great distance. His agreeable talk made us forget his singular appearance. A long loose coat of yellow nankeen cloth—stained all over with the juice of plants, nankeen waistcoat [with enormous pockets, buttoned to the chin] over a pair of tight pantaloons. His beard was long, his lank black hair hung loosely over his shoulders. . . . His forehead was so broad and prominent that any tyro in Phrenology would instantly have pronounced it the residence of a mind of strong powers. . . . His words afforded a sense of rigid truth, and as he directed the conversation to the study of the natural sciences, I listened with as much delight as Telemachus could have done to Mentor. I laid my portfolios before him. He turned to the drawing of a plant quite new to him, inspected it closely, shook his head and told me no such plant existed in nature. I told my guest the plant was common in the immediate neighborhood. He importuned: "Let us go now." We reached the river bank and I pointed to the plant. I turned to Rafinesque and thought he had gone mad. He began plucking the plants one after the other, danced, hugged me, told me exultingly that he had not had now merely a new species but a new genus."[3]

This firsthand account truly revealed Rafinesque's deep passion for nature, as well as his ecstasy upon finding a new plant that he could name. Eaton, Torrey, and others had also been astonished at Rafinesque's naming mania. Public revelation of holes in one's socks would embarrass all but Rafinesque, whose childlike lack of regard for the niceties of appearance contributed to his label as "eccentric." Rafinesque carried no luggage except his specimens, and though offered clothing, he declined.

According to Audubon, Rafinesque's scientific conversation was impressive, and the visitor was "well fitted" to give him advice. At dinner he surprised the Audubons by his charming and delightful conversation. The first evening, the household was awakened by a commotion, when Rafinesque, a guest in the house, was found running naked around his bedroom, strewn with plants, trying to swat a bat with Audubon's violin, and in doing so smashed it ("my demolished Cremona"). Cooly, Audubon writes that he later procured bats for his guest's inspection and classification.

At Rafinesque's request they explored the wilds around Audubon's home, through thickets of cane twelve- to thirty-feet high that one could barely penetrate. Led by Audubon on their first walk, they were about to go around a fallen tree "when out of the center of the tangled mass of branches forth rushed a bear, with such force, and snuffing the air in so frightful a manner" that Rafinesque was terrified. In his attempt to run, he fell and was pinioned between stalks of cane. Despite his thorough fright, Audubon could not refrain "from laughing at the ridiculous exhibition which he made." They continued on, and the way became more and more tangled. "I kept my companion in such constant difficulties, that he now panted, perspired, and seemed almost overcome by fatigue. . . . The thunder began to rumble . . . heavy rains drenched us . . . briars had scratched us, nettles stung us." Rafinesque's serious doubts about surviving the ordeal were countered by Audubon's pontifications on courage and hope.

"Rafinesque threw away all his plants, emptied his pockets of fungi, lichens and mosses. I led him first one way, then another until I myself, though well acquainted with the brake, was all but lost in it. I kept him stumbling and crawling on his hands and knees, until long after midday." In fact, Rafinesque was deliberately misled, for they were close to a river where without difficulty or delay, a boat could be hailed. William Swainson, Rafinesque's reputed friend from his days in Sicily, seemed to regard his encounters with Audubon as "laughable,"[4] and Amos Binney, who despised Rafinesque, found Audubon's story "a very amusing account,"[5] while a sympathetic biographer, Call, considered Rafinesque victimized and Audubon's hoax "cruel and reprehensible."[6]

This sadistic exercise took on another form. Audubon drew and gravely described as many as fourteen fantastic fish (and some birds) of incredible size and color, chimeras that existed only in his imagination, and all were deliberately created to deceive. He showed the drawings to Rafinesque, who unsuspectingly classified them and included them in his classic work on the fish of the Ohio River. A serious work was marred, American zoologists were confounded, and Rafinesque's reputation, already questionable, suffered further. Audubon probably destroyed the drawings of these fish after their purpose had been served. Some of these fish were so strange, one wonders how Rafinesque could have been a victim to such an outrageous hoax. One mythical fish, the "Devil-Jack Diamond fish, *Litholepsis admantinus*," was between four and ten feet long, weighed up to four hundred pounds, and was covered with "stone scales" up to one inch in diameter, that were "ballproof. . . . It lies sometimes asleep or motionless on the surface of the water."[7] Rafinesque credited Audubon for the descriptions of the fish and credited others from hearsay.

Rafinesque, an innocent who respected Audubon and admired his splendid work, had no reason to believe that Audubon was so mean-spirited as to make him a buffoon, an object of scorn. Yet after all this, Rafinesque seemed to be unaware of his humiliation and remained unbelievably credulous about the fish. Audubon never used Rafinesque's name in this adventure, only a thinly disguised "M. de T." But how he could not recognize himself as the victim and how this seasoned naturalist could be misled by such ridiculous accounts is puzzling.[8] Quite simply, neither Rafinesque, whose credulity knew no bounds, nor science were served by this joke.[9]

Although Audubon had treated his guest roughly, and thought him "crazy," he admired him as a "great field man," (his favorite kind of naturalist) and called him "renowned," and "a most agreeable and intelligent companion." After Rafinesque had examined his work, Audubon wrote, "[H]is criticisms were of the greatest advantage to me, for, being well acquainted with books as well as with nature, he was well fitted to give me advice." Yet, Audubon's widely read description of Rafinesque reinforced the image of Rafinesque as unreliable, childish, and ridiculous, confirming the judgment already held by the Eastern establishment—eccentricity was translated into unreliability. On balance, the benefit derived from Audubon's praise was far less than the harm incurred by his "humorous" anecdotes about poor Rafinesque.

To this writer, there is little question that Audubon's well-known, colorful stories were nothing less than malicious. Perhaps Audubon, who so wanted to pass as a "real" American, saw an embarrassing version of himself in his visitor—a foreigner, and a Frenchman at that! His calm *American* demeanor contrasted with his guest's wildly excited escapade with the bat. His account sets Rafinesque's bumbling incompetence against his own cool *American* effectiveness. Credulity is strained by Audubon's self-proclaimed equanimity, as his precious Cremona violin, surely a treasure on the frontier, was smashed to bits (possibly a gross exaggeration). Audubon, no stranger to self-serving behavior and with a reputation for bending the truth, was distancing himself from this foreigner by contrasting his own utter acceptability with that of the peculiar, "different" stranger. If any explanation for this schadenfreude can be divined, its origin could lie in Audubon's sense of helplessness and frustration at presiding over a failing business in Henderson and seeing his possessions destroyed. Yet, the affair remains puzzling, and Rafinesque's lack of resentment would appear to be quite inexplicable. Perhaps the events never took place; the story might have been created by Audubon when writing his memoirs as a kind of private joke between the two.

After three weeks of Audubon's hospitality, during which he collected plants, shells, bats and fish, Rafinesque did not appear for tea one evening and was not heard from for almost a month—he had cleaned out his room and had disappeared. Audubon expressed concern that he had perished somewhere in a swamp, but several weeks later Rafinesque sent a polite letter of thanks without any explanation for his sudden departure. Possibly, Rafinesque may have slowly realized that he had been had and was unable to bear any more of Audubon's patronizing regard and amiable malice. In his *Life of Travels,* he barely mentions his visit to Audubon, though in other accounts he calls Audubon "my friend." There are several instances in which Rafinesque was willing to wipe out what surely must have been old grievances and to renew former relationships. It is odd that after this encounter there seemed to be no hard feelings on either side. Rafinesque reviewed Audubon's *Ornithological Biography* favorably and is said to have named a part of a cave near Lexington that was infested with bats *Audubon Avenue,* while Audubon invited Rafinesque to accompany him on an expedition to Florida.

Rafinesque continued his explorations along the Green and Wabash Rivers, the "Barrens of Kentucky" and the "Prairies of Indiana and Illinois."[10] "I often went 10 miles by narrow paths without meeting a house, and nearly lost myself." His meanderings took him through near midwestern pioneering towns with names that evoke images of the early Republic—Gallipoli, Steubenville, Wheeling, Marietta, Shippingsport, Troy, Evansville, Yellow Bank, Morgantown, Hardinsburg, Shepherdsville, and Frankfort. Part of his journey was made in the company of an itinerant peddler of clocks.

In his wanderings Rafinesque came upon *Harmonie* in Indiana, a theocratic, communistic society on the lower Wabash River that he found worthy of study, for it might possibly be a place of refuge. The Rappite colony had been founded by George Rapp, who ruled by Divine Right and each day assigned tasks that the Lord had imparted to him the night before.[11] Being in the region, Rafinesque appeared there one day "on foot, with a bundle of plants under which a peddler might groan." Here Rafinesque made the acquaintance of Dr. John Christoph Muller, bandmaster of the colony, who possessed a fine herbarium and filled notebooks with observations of plants throughout their blossoming, budding, and foliation stages—very early studies in developmental biology. The value of this work was immediately apparent to Rafinesque, for change and transformation in time were integral parts of his thinking, whether it was of culture, language, or animal and plant forms. He urged a correspondent, the physician-naturalist Charles W. Short, to use Muller's dynamic approach as a model and to follow Muller's lead.[12]

Rafinesque considered his midwestern expedition extremely successful, having collected, described, and classified fossils, plants, shells, bats, rats, mice, reptiles, fish, and insects at an alarming rate. Crossing the Wabash River and traveling as far as the "Prairies of Illinois," he had collected six hundred species of plants of which, in his judgment, sixty were new.[13] One kind of bat he had collected in Henderson was thereafter called Rafinesque's bat, *Nycticeius humeralis.* Rafinesque walked 115 miles from New Harmony to Louisville, where he shipped many of his specimens to a temporary home base in Pittsburgh, while others, in keeping with his desire for free exchange among naturalists, were dispatched to a network of friends, including his companion from Sicily, William Swainson.

Three weeks were spent in Lexington with his good friend and patron from Italy and Philadelphia, John D. Clifford. Clifford, from a Philadelphia merchant family, had moved to Kentucky to establish a factory and become Lexington's leading citizen and maintained financial interests in several cities. Like Rafinesque, a serious collector and scholarly botanist, he had formed a private museum of fossils and plants, which Rafinesque studied and drew, taking great pleasure recognizing a number of new genera and species.[14] Clifford enjoyed Rafinesque's company, especially on collecting trips, and he induced him to settle in Lexington by procuring a professorship for Rafinesque at Transylvania University. Rafinesque's future looked bright; he was protected by a generous patron for whom he had a genuine affection, and the possibilities for collecting materials with his friend were limitless. He was pleased to "settle in a healthy and pleasant town," from which he could investigate the immense, bountiful Mississippi valley, an unexplored land on his map, and he had a prestigious position at an institution that was the first of its kind west of the Allegheny mountains. This was one of the few happy times in Rafinesque's life, but before he settled in Kentucky he had to return to the East to settle his affairs.

Rafinesque traversed the Alleghenies on foot and by wagon a second time, collecting all the way. At Chilicothe, Ohio, "I saw the first great monuments and pyramids or altars, of the nations of N. America; they struck me with astonishment and induced me to study them." His interest in the early civilization of America was pursued during his years in Lexington, where he mapped out and described many ancient mounds of Native Americans. In his enthusiastic way, he believed he had discovered the ruins of a great city—the capital of an ancient civilization in the Lexington area. His observations and thoughts on the subject resulted in a steady flow of papers, a series of reports sent to his friend Samuel L. Mitchill at the Lyceum of New York, who duly published them.[15]

After reaching Lancaster, he ended his journey to Philadelphia by coach, where he spent the winter of 1818–19. He had arrived with large numbers of shells and fossils that he intended to sell, believing them to be rare and precious. They aroused great interest, but no one offered to buy them. In typical Rafinesquian assessment of the situation, he thought he should have been able to sell his shells for "5 to 20 guineas a piece in England."[16]

While in Philadelphia, Rafinesque claimed he was offered a professorship, though he did not reveal its source; he also had an offer of a partnership in a trading venture. He declined both because of his previous commitment to Transylvania University and to Clifford. With all of his traveling and fieldwork, Rafinesque was able to publish a remarkable twenty-five papers in 1818 on a scattering of subjects, and in 1819 he published another thirty. Winter in Philadelphia was a busy time for him, corresponding with his colleagues and with societies, domestic and foreign, shipping specimens to his colleagues, including Baron Cuvier in Paris and William Swainson in England among others, attending meetings of the Academy of Natural Sciences, and working in its library and museum—but above all he wrote.

In December 1818, in Philadelphia, Rafinesque proudly issued his grand totals of new genera and species of plants and animals he had discovered in his recent journey: quadrupeds, thirty species; birds, three genera and thirty-eight species; reptiles, thirty species; fish, seventeen genera and sixty-eight species; crustaceans, one genus and six species; insects, forty species; worms, four genera and eight species; polyps (mostly fossilized), twenty-eight genera and 173 species—a total of eighty-one new genera and 621 new species.[17] Despite all his explorations of the most inaccessible places in vast areas of the near West, American naturalists found these remarkable numbers too good to be true.

Chapter 7

KENTUCKY
1819–1826

Whatever may yet be my fate, inaction does not suit me.

Rafinesque's return to Lexington from Philadelphia began with a journey by steamboat to Baltimore. From there he walked to Frederick, Maryland, and Harper's Ferry, followed the Potomac River to the Cumberland, and then crossed the Allegheny Mountains to reach Pittsburgh. Here he delivered his map of the Ohio River Valley, carefully constructed from information gained on his previous trip to the region, to the firm of Messrs. Cramer and Spear, who paid him a disappointing one hundred dollars. He continued his trip on a keelboat, walking part of the time, surveying Indian mounds at Marietta, Ohio, and collecting "many fine plants."[1]

Arriving in Lexington in midsummer, Rafinesque found the university deserted with everyone on vacation, and his patron Clifford resting in the countryside for his health. He sought him out in the hills, and together, for the remainder of the summer, they explored Indian mounds and collected plants, minerals, and fossils. By autumn, Rafinesque was settled in at the university, the first professor of Natural History in the West.

Rafinesque, an odd-looking foreigner, balding and getting fat, who spoke correct English with a strong French accent, had transplanted himself from Sicily to the American frontier where fully formed communities, copies of those in the East, had mushroomed in the wilderness. Another Frenchman, Alexis de Tocqueville, touring America in the 1830s, was impressed by towns and villages in forest clearings that Easterners and European immigrants had created—the fulfillment of a Romantic ideal. Rafinesque's impression must have been similar to that of Tocqueville who wrote: "[T]he whole country is but a forest. I might add that everywhere where a clearing is to be seen, which is rare enough, the clearing is a village. They give to these villages the most celebrated names of ancient or modern cities, such as Troy, Rome, Liverpool, etc., etc. Besides, these burgs need only eight or ten years to become cities, wherever there is a collection of men and a certain number of buildings. The construction of the houses, which are generally of wood, is not lacking in elegance; their style is often imitated from the Greek." Tocqueville noted that the men he met were "carefully dressed, his house is perfectly clean; usually he has his newspaper beside him, and his first concern is to talk politics with you."[2]

On the other hand, Charles Dickens was appalled by what he saw and mockingly named newly minted "cities" along the Mississippi River (really collections of miserable shacks) *Eden,* and *New Thermopylae*, and many people in them were for the most part ignorant and hypocritical, constantly spouting slogans about liberty and freedom but passionately committed to slavery and in a frenzy to make their fortune.

Transylvania University in Lexington, Kentucky, the frontier's premier center of learning, was to be Rafinesque's home for the next seven years.[3] The town had been settled in 1779 and had been incorporated in 1782, a decade before Kentucky joined the Union. Just recently free from attack by displaced Native Americans, Lexington could now boast of a seminary, which by a decree of the Kentucky legislature in 1798 was combined with its rival, the Presbyterian Kentucky Academy, to become Transylvania University. A rather primitive institution, its backers were convinced it would eventually become the Oxford of the West.[4] By 1820, the population of the town was about 5,300 while surrounding Fayette County contained almost 18,000 people, of which more than 40 percent were slaves—only 133 black people were free. This prosperous community was known for its vigorous support of educational and cultural institutions, which included the Lexington Athenaeum, the Union and Whig Philosophical Society, the Medical Society, the Masons, the Harmonic Society, private academies for boys and girls, amateur and professional theatrical companies, circuses, and horse racing. The slave trade was vigorously pursued, though it was said that the treatment of slaves in the locality was "relatively mild and benevolent."[5]

Rafinesque arrived at a time when the town had become the cultural, commercial, and manufacturing center of the West, though commercial dominance soon passed to Cincinnati. Set in a town of stores and churches, with two bookstores and a coffee shop with forty-two newspapers for browsing, citizens expected that the university would soon rival its Eastern predecessors. Though supported by the State, the university was governed by a Board of Trustees that consisted mostly of members of the Presbyterian Church with conservative and fundamentalist leanings. When the State Legislature appointed a new Board of liberal persuasion in 1817, with Henry Clay a member (leader of the national Whig party and presidential candidate) they were able to name Horace Holley president. Despite the active opposition of the Presbyterian faction, who never ceased to harangue Holley with sermons and pamphlets, the school flourished in the years Rafinesque was there. Holley, a sophisticated Unitarian clergyman from Boston, a graduate of Yale, and an overseer at Harvard, was the right man for the job. He was admired by the faculty, students, and townspeople and was handsome, charming, an outstanding orator, very sociable, and political. He was happy to attend dances, the theater, and even the racecourse.

The size of the school increased until its enrollment was as large as those at eastern colleges, and an impressive building was constructed. Academic standards rose, the three faculties—Arts, Law, and Medicine—grew in stature as Holley recruited the best possible people. The medical school, the first in the West and one of the best in the nation, was staffed by five professors, mostly graduates of the University of Pennsylvania.[6] The university could boast a remarkable group of alumni that included the Texas promoter Stephen F. Austin, seventeen congressmen, three governors, six U.S. senators, and Jefferson Davis.[7] During Davis's incarceration after the Civil War, he occupied his time studying botany, an interest probably initiated by Rafinesque years earlier. From the very beginning of the Republic, the dispersion of settlers over the continent was attended by a decentralization of the educational and research systems, which has proven to be one of America's greatest sources of strength.

Rafinesque was pleased with the civilized interests of Lexingtonians, the charm and refinement of the town, the graciousness and lively curiosity of its citizens, and the beauty of the surrounding landscape, bluegrass country reminiscent of Provence. Lexington supported four newspapers and four literary journals, proportionately far greater than Philadelphia, New York, or Boston. According to Rafinesque, all the schools of the West were "mere grammar schools" compared to Transylvania University: "The surname of *Athens of the West* has already been given to this town, and methinks on very plausible and reasonable grounds. There is certainly not a single town west of the mountains that can rival with it on that score. Pittsburgh, Cincinnati and Louisville are mere commercial towns, and all attempts to establish there permanent seats of learning have failed. Let Pittsburgh become the Manchester of the

Western States, Cincinnati their Liverpool and Louisville their Bristol; but Lexington must be their Edinburgh."[8]

While Rafinesque was passionately devoted to the discovery of new species and genera of plants and animals, he was equally devoted to the dissemination of knowledge, and fittingly David Starr Jordan wrote of him, "Rafinesque was the first teacher of Natural Science in the west, and there were not many anywhere in his time."[9] It is to Rafinesque's credit that he was prepared to teach a subject that was not well favored and in a remote part of the United States where few were as qualified as he to present the most advanced ideas of the day. If he had any real craving to be the center of attention, as many of his colleagues insisted, he was now well placed. Rafinesque's aim was to provide an education that would instill in the young a reverence for nature and the biological disciplines, enabling students to comprehend their beauty and utility, for the nation could prosper only if its citizens had a proper appreciation of science and of knowledge in general.

Rafinesque, who did not fit comfortably into the social class system of Lexington, was indulged but not warmly welcomed by his fellow professors, although he was appointed Librarian and Secretary to the Faculty with their approval (both paid positions). He constantly submitted detailed lists of requests and demands to the Board of Trustees. Few at the university had any sympathy or understanding of science and natural history, considering them a threat to their religion and competition for students' attention. The main purpose of the university was to educate ministers, especially for the frontier. Even President Holley, considered a "progressive," was not completely convinced that a professor of natural history, especially a collector, was desirable at Transylvania,[10] but the trustees, goaded by the influential Clifford, a man with a passion for natural history, gave way. As Lexington's most prominent citizen, he had the clout to have his way, and so Rafinesque was hired by the university as professor in the newly created Department of Botany and Natural History, and to add weight to the appointment, his title also included the professorship of modern languages.

The newcomer was adopted by the kindly, urbane Mrs. Mary Austin Holley, wife of the president, for whom he had a tender affection. Rafinesque taught her Italian, helped her adjust to the isolation of life in Kentucky, and enlivened her existence by providing conversation and continental panache—bowing deeply to ladies and kissing their outstretched hands—while she provided this "lonely and friendless man" with warm friendship. When her daughter Harriette was married, Rafinesque wrote and recited his *Epithalamium or Nuptial Ode for the Wedding of Harriot Holley and William Brand.* He signed the printed copy of his work, "Constantine, the Grecian Bard." Mrs. Holley's roses and jasmines prospered when Rafinesque's expert gardening advice was followed, and he in turn credited her with the discovery of *Spirea rosea.* She seemed to have recognized his brilliant, if eccentric, qualities and had a calming effect on him, making sure that his clothing was clean, that he washed his face and combed his hair, and that he ate proper meals—all otherwise neglected when he was in a scientific frenzy. Rafinesque was not in the custom of washing himself regularly, and he had little experience with the pleasures of domestic life. A member of the Holley household remembered: "He [Rafinesque] was a great admirer of Dr. Holly [sic] and came frequently to the house to talk on subjects of interest to him. He was never an inmate of the house although his face was a familiar one there. . . . He wrote verses, English, and Italian and Latin, I think, and brought them to find an audience with us."[11]

Rafinesque was in the habit of sketching visitors and friends. Mrs. Holley's likeness was surrounded by the artist's touching inscription:

> Emblèmes sincères de douceux et beauté
> Ces fleurs sentent dire la simple verité,[12]

Though Rafinesque was adorned with the imposing title, *Professor of Botany, Natural History and Modern Languages*, his threadbare existence continued, for he received no salary, only room and board, candles, and wood for his fireplace, at a time when Holley's salary was $3,000 per year, with a house.[13] Rafinesque was pitilessly exploited by the school authorities, who thought he was so passionately involved in his work that he had no earthly needs, and they even complained about the number of candles used by this tireless writer who worked well into the night. His cash income was to come from whatever he might scratch out, charging students a fee for admission to his lectures—not a lucrative prospect. Professors in the medical school did the same, but they were also on a real salary, and those in the college ranged from $600 to $1,200 per year.[14]

In the fall of 1819, he organized a course of lectures on natural history. Advertised in the local newspaper, the *Kentucky Reporter,* and by handbill,[15] he charged ten dollars to attend all twenty lectures, which covered astronomy, geology, zoology, meteorology, vulcanology, minerals, mines, and fossils. From an outline of the course, attended by about sixteen women and twelve men, it appeared to be broad and well-balanced, reflecting Rafinesque's deep understanding of the natural history and science of his day.[16] Privately, Rafinesque offered courses in French, Italian, and Spanish, as well as courses in English for French students—who paid ten dollars per quarter. All lectures were open to the public (both ladies and gentlemen) and to students of the college and medical school. In the spring of 1820, he gave a comprehensive course of twenty lectures on botany, which included the anatomy, physiology, taxonomy, and practical study of plants. Two lectures were devoted to the Linnaean and the natural system of classification. This arrangement, whereby lectures were given at the university for both students and the interested public, continued throughout his Lexington years, though there were times when courses had to be postponed for lack of takers. Rafinesque would teach anything, including geometry and map drawing, if he could round up paying students.

Fortunately there was a sizeable pool of students; in 1820, 235 students were taught by eight professors and six tutors, and there were also about two hundred medical students.[17] By including lectures on medical botany, of major importance to physicians who often gathered plants believed to be of therapeutic value, Rafinesque attracted medical students to his course, offering a bargain they found hard to resist, and if students enrolled in groups he would give them special rates. To earn a few extra dollars, he announced in the newspapers that he was offering special lectures for fifty cents on topics such as the nature of knowledge, mankind, the human mind, phrenology, craniology, medical botany, and materia medica. He was not a conventional lecturer, and though popular, people really did not know what to make of him.

Because Rafinesque was born in Turkey (Constantinople) of a mother raised in Greece, newspapers referred to him as a native of Greece, well qualified to lecture on the war of Greek independence from Turkey (1821–1828). Sympathy for the Greek cause was strong in America, especially in Kentucky, where Henry Clay was a passionate advocate of the Greek cause. One such lecture, a public event attended by faculty and students as part of a fundraising drive, was reported by Rafinesque in the *Kentucky Reporter* under the alias "Constantine of Byzantium." "The joyful thanks of the Grecian heroes, widows, and orphans await you and will be your reward at the reception of your gifts. Despots may frown but heaven will smile and register the deed."[18]

By 1823, a university administration annoyed by a quarrelsome professor of natural history and botany who was absent from school much of the time, ordered that his salary should cease, but they saw fit to pay him $200 per year as the librarian and keeper of the museum, a position created for him by President Holley. By this time, relations between the overtaxed Holley, who was harassed without mercy by the Presbyterian faction, and Rafinesque were becoming strained, and Holley's "charitable" offering was made to reduce a multitude of Rafinesquian activities (such as the creation of a large municipal garden) to a few specific tasks—lecturing among them. Rafinesque accepted the librarianship, but he had other plans.

Fortunately at this time, his botany course was attended by 108 paying students; his teaching was innovative and lively, capable of drawing a large audience. Rafinesque used living specimens in the classroom to make his points, an early example of a new and improved method of teaching that seemed highly successful. General George W. Jones, who attended Transylvania University from 1821 to 1825, reminisced seventy years later:

> He often lectured to the students in College and in a most entertaining manner to the great delight of his audiences. His lectures on the ants were particularly instructive and interesting, causing many of the students to laugh heartily when he gave us the history of ants, especially when he described them as having lawyers, doctors, generals, and privates, and of their having great battles and of the care by physicians and nurses of the wounded etc etc. . . . I would now give any reasonable sum to hear him repeat one of his lectures. . . . He was a man of peculiar habits and very eccentric, but was to me one of the most interesting men I have ever known.[19]

A young woman who had observed him in and out of the classroom recalled that "His classroom was the scene of the most free and easy behaviour . . . a most eccentric person, his extreme absent-mindedness contributing to his foreign ways to make him peculiar. He went into Society while in Lexington, and was a good dancer but had no conversation save on his favorite topic."[20] An acquaintance wrote of him that he was "a small peculiar Italian with a large rather bald head and stooping figure . . . with many peculiarities and not much dignity." And another described him as being "careless in his style of dressing, indeed, his clothes never fitted him and appeared to have been made for some one else, and he got them by accident. I think he was not a cheerful man. I have no recollection of having ever seen him enjoy a hearty laugh. He was an eccentric man."[21]

Rafinesque captured the imagination of his students and his peers, who had never seen anyone like him. He was a "personality" who probably exemplified to them the "high" culture of Europe, especially attractive to those who were aware of the inadequacies of their own provincial lives on the American frontier. Rafinesque's students could hardly escape exposure to his protean interests and enthusiasms. He held strong, informed views on many subjects, which he could buttress with scientific information and reasoning. One such view was his belief that "atmospheric dust" was constantly being deposited on the earth and was responsible for the burial of ancient monuments.[22] Of special interest to students was his discussion of the geology and physical geography of locales with which they were familiar—the Ohio and Mississippi River valleys and the Allegheny Mountains.[23]

He was appreciative of women, and at this early date they responded by flocking to his classes. From his lecture *Picture of Woman,* it is apparent he believed women could learn as well as men without harm to their delicate constitutions.[24] Women's assistance in collecting plants was welcomed and freely acknowledged.[25] He encouraged their friendship and he enjoyed dancing with them, but one senses that intimacy was rare or nonexistent; to women he was merely an odd amusement and his relationships were most probably Platonic.

Rafinesque fancied himself a talented artist who could draw and paint the plants he collected, but he wrote that he "seldom availed myself of my talent." In reality, he was less than mediocre, and what he lacked in talent he made up for in confidence. He drew portraits of many ladies, including his mother as he remembered her through the haze of twenty years, and his sister when she had been fourteen years of age. Under each portrait he would insert a caption, often coy, sometimes tender, and other times regretful of the loss of pristine virtues. One sketch of an unknown "Juliet" was accompanied by:

> I knew her in the prime of her beauty and youth:
> When she was the chaste emblem of candor and truth
> But alas! what a change!
>
> Another portrait was graced with:
> Elle était séduisante
> Belle aimable et charmante.[26]

He sang in hope and praise for women in Whitman's declamatory style, though his overheated song was a bit clumsy and a little importuning: "And you Women! female angels of this Earth, budding and blooming awhile to please and delight, maturing to renovate and adorn mankind; you are blest among all by the gifts of loveliness, sentiments, compassion and mildness—do not misuse the power of your charms and examples, but exert them ever to redeem the miseries and crimes of those who owe you birth and love."[27]

Throughout 1818 and until April 1819, prior to moving West, Rafinesque published scores of papers in the New York–based *American Monthly and Criti-*

cal Review, usually one paper in every issue—on botany, zoology, meteorology, and geology. Despite his isolation in Kentucky, his removal from important editors, and his exclusion from a journal such as Silliman's *American Journal of Science*, the torrent of communications continued. Publication in the *American Journal of Science* came to a sudden halt when its editor, Benjamin Silliman, fearing this inundation, returned a bundle containing twelve papers. He was also responding to stories that cast doubt on Rafinesque's integrity. Thwarted in America, Rafinesque's work appeared in the Belgian *Annals of Physical Sciences* through the influence of his friend Bory St. Vincent. He wrote progress reports on American science in this journal, which informed Europeans of what was going on in the New World. Sending his work to France and Germany, he published five papers in 1819 and twenty-nine in 1821.

His ultimate solution was to establish his own journal in which he could, without hindrance, provide the world with his views and with endless amounts of information. In the meantime his papers appeared in the Lexington-based *The Western Review and Miscellaneous Magazine*—eight in 1819, and twenty-one in 1820 (the last year of its publication). The *Western Review* had been a learned and scholarly voice in the wilderness emanating from Lexington, which for whatever reason was unable to resist the furious submission of Rafinesque's manuscripts. He published his poetry; "sacred odes"; lengthy book reviews; works on geology, astronomy, meteorology, botany, the fish of the Ohio River, statistics on education in America, and the archaeology, history, and language of the Native American. The journal published his articles on salivation of the horse, the oil of the pumpkin seed, the galaxy or Milky Way, and a vast display of other arcane knowledge.[28]

In May 1820, John D. Clifford, Rafinesque's closest friend, patron, and mentor, died suddenly at the age of forty-two. The two men had been preparing for one of their collecting expeditions in western Kentucky and Arkansas, traveling about in Clifford's carriage, when Clifford "was seized with a fit of gout in the stomach, which proved fatal in a few days." Rafinesque was distraught for he had a stronger affection for Clifford than he had for any other man. "This loss of an intimate and zealous friend was blasting to all my hopes and views."[29] The major stabilizing force in Rafinesque's life was gone, and whatever measure of financial security he enjoyed suddenly disappeared. By 1822, without Clifford, Rafinesque's status at the university became precarious, and for the remaining eighteen years of his life he met with repeated failure and disappointment—the underlying cause of bizarre and even psychotic behavior that manifested from time to time.

On the death of his friend, he wrote a deeply felt elegy in Italian, which appeared in *The Western Review and Miscellaneous Magazine*, and over time Clifford's fine museum collection of artifacts, stones and ceramics, and plants were scattered, including his "Triune idol," a stone sculpture that purportedly was strong evidence for a Hindu presence in early American history. Rafinesque considered returning to the East, but he felt he could not bear the effort and expense of moving his library and collections, and he was reluctant to leave a part of the country he wanted to explore collecting its flora and fauna.

Rafinesque's urgent desire to bring Culture and Enlightenment to the West found expression when he began publishing *Western Minerva,* or *American Annals of Knowledge and Literature,* a quarterly journal costing two dollars per year, paid in advance. The publication was completely under the control of its editor and sole contributor, Constantine Samuel Rafinesque, who was also the journal's underwriter—a formidable burden for this penniless man. The journal was to contain "Original Essays upon Science, the Arts, Literature, and subjects connected with the Civil and Natural History of the Western States." The lack of subscribers outraged Rafinesque. He claimed that his journal did not attract readers because it was "too learned and too liberal," in contrast to popular magazines and literary journals "which contain hardly anything beyond plagiarisms and vapid trash." For a single number of the *Western Minerva,* Rafinesque wrote over fifty articles, poems, and musings, signing them in many different ways, perhaps reflecting multiple personalities whose identities were feebly concealed—his full name, C.S.R., Mentor, Constantine for poems in English, Costantino for his Italian poems, and other poems by Delia, Sweetlips, Eleonora, and Oscar. These creations were entitled *Lines to Maria, Who asked me whether I should like to Love in a Cottage, To Silvia, My heart is gone, A Melody, The Man I'll Love.*

For balance, Rafinesque included learned papers on legislation, principles of political wisdom, moral philosophy, metaphysics, works of Benjamin Franklin, meteorology, botany, zoology, manufacturing, agriculture, education, philology, and fine arts. Rafinesque valued the fact-gathering approach in demographic studies, and while in Sicily he had collaborated with Giuseppe Emmanuele Ortolani in census work. In the United States he assembled valuable quantitative information about Lexington, which included the number of dwellings, churches, factories, and other buildings; the total population; and a breakdown of the population by trade and professions.[30] Rafinesque also wrote about inventions, some his own, such as a "cubometer," by which he could determine the "bulk" of any object by immersing it in water—surely a *reinvention.* He could not be labeled a gadgeteer, but in the course of his career he wrote in earnest of "inventions" for fireproof architecture, aquatic railways, steam ploughs, a method for navigating in shallow waters, and the production of artificial leather.

A semi-mystical paper on female freemasonry described the reception or initiation into a Masonic-like cult, a description of which was "translated from the hieroglyphic language, used in the female lodges of Germany, Denmark and France . . . the translation was by a holy brother of the highest degree. The candidate is conducted into a dark room, and left there alone while the lodge is forming. . . . She is introduced blindfold in the lodge."[31] The scene could have been taken from *The Magic Flute,* with its secular priests presiding over rites introducing the unenlightened to the world of Nature, Reason, and Wisdom.

Even more peculiar was a paper on *polygriphs*, one of an army of words coined by Rafinesque that pertained to peculiar enigmas or riddles, which were popular at the time. He proclaimed flirtatiously, that they would delight the ladies, something he said he always tried to do. This bonbon, sometimes written in a fanciful archaic style, was a mumbo-jumbo of plays on words, numerology, and magic formulas that probably had its origin in the Cabbala. It makes little sense today, and most probably meant little more than 170 years ago.

The *Western Minerva* was an unrestrained work of pure, unedited Rafinesque. The freemasonry article was a barely concealed challenge to women to do better, to live up to their capabilities for great achievement and fulfillment—emblematic of the admiration and goodwill he had for them. They inspired the deeply romantic Rafinesque to write coy poetry in the style of Robert Herrick, almost all of which was gallantly addressed to "my fair readers" or "fair ladies," and all of it rather ordinary, as were his characterless portraits of women who all seemed to look alike. He included twelve poems in his article on polygriphs, and each was given a name and was dedicated to a young woman:

No. 11 A Calligrograph—To Miss H.D.

I fly like birds, and when I die I weep.
Change my two last limbs, and I smell like a pink,
Change all my limbs, except my head and heart,
And you will find, besides a happy sign,
The emblem of kingly and noble power,
What we trust in, and yet what we despise.

No. 12 A Calligrograph.—To the lady of my Thoughts.

By love inspired, I dream of love and you
Each night and day: waiting the happy hour,
When your dear self may my subject become.
Change all my limbs, but keep my head and heart;
This last is yours you know, no longer mine:
Then let me prove, what I hope to receive
From your beloved self.
If your heart shall guess my meaning,
My wish shall be your wish,
And we shall hasten the time
Of your change and my own,
Of my joys by your own.
CONSTANTINE

The *Western Minerva* was designed to appeal to both sexes, a kind of family magazine in which there was something for everyone, but in the end Rafinesque's efforts were appreciated by no one. Two hundred and fifty subscriptions were needed for the journal to pay for itself but there were only one hundred. Only one number was published, and the printer destroyed all but three copies because he was not paid.[32] Rafinesque hinted darkly that his enemies had "suppressed" the work by buying off the printer, because he, Rafinesque, knew more than those who would "control the press and crush knowledge."[33] He complained bitterly to his friend John Torrey about the injustice of it all,[34] and he responded to his critics with lively invective.[35]

As one journal ceased publication, Rafinesque would begin with another, a practice followed by many other entrepreneurial journalists. After the death of the *Western Minerva*, he published sixteen botanical and zoological works in the *Kentucky Gazette.* In 1823, a silence of one year followed in which no journal articles appeared because he was preoccupied with establishing a great civic garden in Lexington, and was preparing the ground for his studies on archaeology, ethnology, and linguistics. These works were promised in a series of articles in *The Cincinnati Literary Gazette* and in pamphlets.[36]

A series of papers by Rafinesque on the ancient civilizations of America was begun in 1824. The first paper, an overview of the subject, was introduced by the editor with the disclaimer: "It is very probable however that the grounds on which his opinions are founded might not appear as substantial to most readers as to the learned professor." Why did the editor see fit to publish the work at all? Rafinesque's astounding performance, the first of several in this field, must have had readers shaking their heads in disbelief. Rafinesque claimed that he was familiar with 1,830 monuments, excluding mounds and graves, in 505 sites in Ohio and Kentucky. These had been erected, he said, by three races of people that had inhabited America in succession. The first came from the East and sprang from five North African nations—the Atlantes, Pallis, Warbars, Darans, and Corans—and from five European nations—the Celts, Cantabrians, Cimbrians, Pelasgians, and Tubalans. These were known by the distinctive circular shape of the monuments they erected, accounting for half of the present Native American population, which had divided into one thousand nations. The "second race of men came from Asia by the West," and from it sprang "600 nations and tribes such as the Toltecas, Mexican Natchez, Osages, Chicasaws &c." Their monuments were recognized by their angular shapes, and the third race, from Siberia, was the most recent and was identified by the "rude structure" of their monuments. They separated into six nations—Lenapians, Mengwers, Edluans, Rumsens, Euslens, and Karatits—from which arose four hundred tribes. Rafinesque claimed he had examined the languages of these groups and had reduced them to ten mother languages, all spoken or extinct. These he could reduce "to a single Primitive language, divided in three branches: Iranic, Atlantic, and Scythic." Twenty-five American languages existed, all of which he named, that had arisen from two thousand dialects. One can only wonder about this excessively informed work—where did such information come from?

Rafinesque had a dynamic, evolutionary conception of language and dialects, identifying a Babel of them—all specified, with relationships ascertained. Through his work he claimed that "several great historical problems have been solved": Who were the first inhabitants of America? Who were the ancestors of the Mexican people? And who built the ancient monuments?[37] Rafinesque was remarkably facile in absorbing detail, digesting it, and divining relationships, from which narratives were generated, with the logical and seamless nature of the narratives themselves serving as evidence for their authenticity.

By listing and briefly describing ancient monuments, Rafinesque was a pioneer in establishing system and order in a study of scattered curiosities. He stressed that the history of America did not begin with Columbus, and he was prescient in his plea for the detailed study of languages as a key to understanding the flow of history, the migration of tribes, and the relationships of various human populations.[38] But the flood of his overblown claims elicited reactions of scornful disbelief, for he created whole systems of knowledge upon which rested the most preposterous claims:

Far from entertaining any bad feelings against those who may correct my statements or follow my steps, I invite additional or more correct information. I am prepared to render them all the justice that their labours shall deserve, and I only ask that equal justice should be rendered to me, such justice as is often denied by a few illiberal minds, holding a presumptuous rank among the votaries of Nature and of knowledge.[39]

In one paper, for example, Rafinesque expounded on an American Solomon, Nazahual, the tenth king of Anahuac (Mexico), who cultivated and improved religion, philosophy, science, arts, and literature, and was a great warrior, legislator, reformer, naturalist, poet, and astronomer, and who "exceeded the Asiatic Solomon in many things."[40] A taunting response by a reader, "*B*," probably typified the public's scornful attitude toward Rafinesque:

To C.S. Rafinesque, D.P. &c. &c. Modern Catesby, P.B.T.U.D.K.&c.

Sir,—

No doubt is entertained of the correctness of your statement, in saying that this *American Solomon*, was a greater man than the *Asiatic Solomon:* indeed this is fully proved by his having *caused paintings to be made of all* the STARS, ANIMALS, *and* PLANTS *in Anahuac,*- a devotion to natural history, that did not mark the character of the *Old Bible Solomon.* I am sorry, however, to inform you that some persons in this city affect to doubt whether this *Big Solomon* of yours was in reality a DEIST, as you have asserted: others declare that *His temple* which you say was *nine* stories in height, was but *eight and three quarters:* and I am sorrowful to tell you that I have met with one or two persons, so incredulous and obstinately perverse, as to declare a total *disbelief* in the existence of any such man as you have described, except in your fertile imagination.

Now to settle this matter, will you, my good sir, be so kind as to furnish for the Literary Gazette, your authorities for the statements about the "*American Solomon.*" If you knew "King Nazahual *personally,* and have made your sketch from actual observation?"[41]

Rafinesque was barely able to conceal his annoyance: "I have been called upon, to give my authorities for the Biographical sketch of Nazahual the first: although the demand was anonymous and indecorous, therefore unworthy of notice; since it has been admitted to your pages, it requires a short answer."[42]

The most reasonable explanation for Rafinesque's grand outpouring is that he was under a lot of stress and was suffering from a serious manic disorder that left him unable to control his imagination, which fed on an immense erudition. What Rafinesque believed to be so became so, unable as he was to distinguish verifiable information from the "facts" that emanated from his intuition and fancy, dismissing his critics as fools and enemies who simply did not appreciate a superior intellect. What a trial Rafinesque was for his contemporaries, who had to deal with such an earnest, articulate advocate of nonsense, who at the same time was capable of presenting valuable information and brilliant ideas. Despite some of his preposterous beliefs, he considered himself a serious, rational scientist, intolerant of a pseudoscience such as phrenology.[43] If he had qualified his statements or had cast his stories as mythology rooted in history, in the style of Tolkien or of Frazer's *The Golden Bough* he might have had a kind of legitimacy, and perhaps could have been credited with reinventing a literary genre.

The Cincinnati Literary Gazette had in Rafinesque a knowledgeable and articulate historian, philologist, and grammarian. A scholarly review of a book on Hebrew grammar by the reverend Martin Ruter revealed Rafinesque's impressive understanding of the origin and structure of the language and the relationship between kindred languages. Generous with praise but condescending in style, he pointed out a few errors in the author's understanding of Hebrew and the sources of his error, and suggested a few possible improvements in future editions. Disputing a statement that "the Hebrew has a higher claim to antiquity than any other language now existing," Rafinesque asserted: "We know at least fifty languages as old or older than the Hebrew of the Bible." A Faustian character, he was certain he could know and understand everything, and if he sometimes weakly professed ignorance, it was *sotto voce*.[44] The review ended jauntily with advice (as if to a schoolboy): "We now dismiss the author with our best wishes for his first attempt, and a confident hope that if he applies himself to oriental literature in general, he will be able to enlarge his views and usefulness in this philological department."

In 1825, Rafinesque published a revealing paper on *Useful Inventions* that exulted in the fact that Americans lived in an era of progress in which even "greater discoveries are yet to be made in every branch of knowledge." While "in Europe, Scientific discoveries are more highly valued than mechanical inventions," practical, "lucrative inventions" counted most in the United States. He concluded that "those who have applied themselves, in America, to improve any Science, have never met with an adequate reward, as yet: while mere mechanics have made fortunes by some modification of manual labor." Rafinesque identified himself with those who suffered from inadequately rewarded scientific study, and immodestly declared: "After having made a multitude of discoveries, in many physical, historical and philological sciences, which have been well received in Europe but have attracted no attention or been perverted, in this neighborhood, I HAVE at last deemed it necessary to conform myself to the local sentiments on this subject, as a means to facilitate my ulterior labors, and to become useful to myself, and others like me. . . . I HAVE THUS SUCCEEDED to achieve several INVENTIONS, of the most EXTRAORDINARY NATURE, magnitude, importance and utility." There were three that were "so singular," he felt compelled to announce them to the world at the appropriate time but he would divulge no details at the moment. However, at the appropriate time, he would reveal all, hinting that he was compelled to take this course to thwart his foes. The first invention would make many people rich, the second would "prevent and suppress Vices and Crimes," and the third, "and the most extraordinary, will have for its aim to prevent Wars or attacks, by rendering them so dangerous that none but madmen will attempt them." His dreaded invention that would keep the peace he called a "PEACE-ENGINE." Shortly thereafter, a mocking plea to Rafinesque from reader "Y" to reveal his secrets for the benefit of all mankind, appeared in *The Cincinnati Literary Gazette*.[45] There was always a wide gap between what he promised and what he delivered, but with these inventions, he seems to have gone too far, and appeared quite mad.

In a utilitarian age, Rafinesque always had practical applications of his knowledge in mind, especially in agriculture (cultivation of grapes for the production of wine and mulberry bushes for raising silkworms).[46] Since Rafinesque paid welcomed visits to estates and farms throughout the land and was always generous with expert advice, he probably had a large, although unacknowledged, influence on agricultural practice.

He was constantly inventing and seeking patents, which were ignored by his competition. After his monetary and banking inventions had been expropriated by others without compensation he complained: "Happily I had kept secret and in reserve, several other discoveries of mine. Perceiving this disposition to appropriate my labors and knowledge, I was compelled to foil this kind of swindling or knavery by not taking any more patents; but using secretly my Inventions. Some envious hearts may have blamed me for it; they are probably those who would have been the first to steal them if published."[47]

Besides writing on historical matters, Rafinesque spent much of his time amassing a huge herbarium, and publishing works on botany. At the same time he kept up a lively correspondence with William Swainson, his naturalist friend from Sicily,[48] and several other leading European natural historians including Georges Cuvier, Augustin Pyramus de Candolle, and Sir Joseph Banks. To Cuvier he sent a description of many new genera of animals, which Cuvier published in the *Journal des Physiques*. In 1818 Cuvier had sent a letter complimenting Rafinesque on his discovery of several new species and genera of animals, admitting that some animals he had described had actually been discovered earlier by Rafinesque.[49]

In America, Rafinesque corresponded with Jefferson, but most of his communications were with Zaccheus Collins, the botanist John Torrey, and Charles Wilkins Short, a Penn medical graduate and a dedicated, part-time botanist who later settled in Kentucky and became a professor at the Transylvania Medical School. As a mentor, Rafinesque wrote detailed letters to Short, answering all his questions,[50] urging him to keep a journal and to carefully record his observations on the growth of plants.[51] In his ceaseless wanderings through Tennessee and Kentucky, Rafinesque collected, lectured, and paid frequent visits to his prominent acquaintances—Governor Shelby, General Harrison, and Henry Clay, who always welcomed him for short stays.

Firmly believing that no civilized society should be without a garden, Rafinesque dedicated himself to the establishment of a university-associated civic garden in Lexington. Gardens (and greenhouses) were not only status symbols for aspiring communities and for individuals who wanted to display their wealth, they could also be useful to farmers and physicians and promote the study of plants. In some instances gardens were designed as *tableaux vivants* to reproduce sentimental genre scenes so favored by painters. He was no stranger to formal gardens, and without doubt his thinking about them was shaped by the magnificent gardens of Italy and Marseille, as well as the Hamilton and Bartram gardens in Philadelphia.

Working with influential friends, he induced the Senate of the Kentucky Legislature at Frankfort to issue a charter for the establishment of a botanical garden to be supported, at least in part, by the state.[52] His scheme seemed somewhat unrealistic, for how could a relatively small, frontier town support the grand work he envisaged? When the charter was not approved by the House, Rafinesque and his friends responded by establishing a joint stock company, the Transylvania Botanical Garden Company. For the venture to be a success, a sale of one hundred shares was necessary, and to this end persons living within a forty-mile radius of Lexington were asked to buy shares, but the response was not promising. Rafinesque bought five shares for $250 and Henry Clay bought two shares. However, Rafinesque persisted, and ten acres of land were purchased in Lexington for $1,000. Rafinesque was promoted to superintendent in order to put the garden in "handsome operation." In March 1825 a gardener was hired, tools were bought, the garden was laid out, and trees were planted.

The Board of Trustees of the university graciously accepted Rafinesque's collections of plants, minerals, and fossils, and appointed him curator of their museum, but the offer came to nothing when the university was called upon to properly house the collection. He had amassed more than 37,000 plant specimens in almost 13,000 species, but the entire project came to a halt as financial support withered. Then Rafinesque, the sustaining force of the scheme, left for his collecting trip that summer. Without his constant supervision and promotion, his projects became hopelessly bogged down by the inertia of the populace and the opposition of the "foes of science." Frustrated, he wrote an unhappy letter to his friend, Zaccheus Collins, in Philadelphia admitting failure. In despair he wrote of his inability to make a living or to get his work published.[53]

As an autodidact without wealth or tangible credentials who had attended no schools, he saw value in collecting memberships in scientific societies and in being awarded diplomas and degrees, and each was carefully added to a growing list that was printed on the title pages of his books and pamphlets. Through hard work and persistence Rafinesque achieved positions of some tenuous influence; he was Secretary of the Faculty and the Librarian of the University. A master's degree from Transylvania University was initially denied him because he "had not studied Greek in a College! altho' I knew more languages than all the American Colleges united," but eventually, through persistence, the degree was granted. This mark of distinction was now added to his first honorary membership and diploma that had come in 1814 from the Academy of Natural Sciences of Naples. Then an honorary doctorate was conferred by the ancient and learned society *Natura Curiosorum* of Bonn, called the Imperial Academy of Bonn, which granted him the title *Catesbaeus* or "Dr. Catesby," after the pioneering English naturalist who wrote of his visit to America in the early eighteenth century. Rafinesque was not above calling himself a doctor of philosophy long before he had such a degree; he felt his erudition warranted the title. Diplomas and scientific honors—about fifteen in all—were conferred by Zurich, Vienna, Brussels, Paris, Philadelphia, and Cincinnati, "which have all been expensive rather than profitable honors."[54] Rafinesque was ridiculed for flaunting his degrees and honors—it was just not done in respectable circles.

He had been a lecturer on medical botany and later the author of a book on the subject,[55] and so he had hoped to be the professor of Materia Medica on the medical faculty, a chair that paid well year after year. If the Board of Trustees offered him the chair, Rafinesque was willing to "gradually" donate $50,000 to establish a botanical and medical garden and museum in Lexington. For this position he was required to have a doctorate in medicine, but this was denied him, "because I would not assist to anatomical dissections for which I entertain a dislike." Where Rafinesque could raise $50,000 on his salary remains a mystery; in his desperation, he was enslaving himself by pledging his lifetime salary. He also had hoped to become the professor of Public Economy in the Law School, but this opportunity also vanished. Rafinesque, at an all-time low, was prepared to leave Lexington.

As a treasure house to be explored, Kentucky and the West were alluring, but Rafinesque was as badly out of place as a European *philosophe*. Despite his praise of Lexington, he would have been glad to leave the town for the East as early as September 1819, not long after his arrival, because he soon realized that "the West is not yet mature for Sciences."[56] Upon hearing that Thomas Jefferson, who had retired to *Monticello,* was assembling a faculty for the recently founded University of Virginia, Rafinesque applied for a professorship of Natural History. The inducements he offered if he was hired betrayed his underlying desperation—he would be willing to donate his extensive herbarium and his library to the university, he would spend up to one-third of his salary on books for the university library, and after John D. Clifford died he offered the University of Virginia Clifford's prized museum of fossils and Indian artifacts, which Rafinesque planned to purchase for two thousand dollars.[57] Two months later Jefferson replied, pleading illness for the delay, informing Rafinesque that the Board of Trustees of the university had suspended the hiring of staff for at least one year.[58] The Board, while not a fiction, was obviously a device for diffusing this unpleasant decision. Rafinesque persisted, writing Jefferson friendly, informative letters, sometimes containing seeds, and always inquiring about the professorship. He wrote of his troubles in Kentucky and of the "censorship" of his journal, *Western Minerva.* He sent Jefferson his published works as they appeared and testimonial letters from Baron Cuvier, William Swainson, and Augustin Pyramus de Candolle. In addition, long lists of European and American scientists were provided as references. Rafinesque offered to establish a garden at the university at his own expense. A discussion of the problem of hiring suitable professors and the necessity of looking to Europe for candidates served as an introduction to advancing his own candidacy once again: "I do not know a single Individual either in the U. St. or in Europe, who is *at the same time equally acquainted* with Geology, Mineralogy, Meteorology, Zoology and Botany as I am."[59] He was probably correct in his assertion.

Rafinesque kept pressing while Jefferson politely evaded all offers. In an attempt to engage Jefferson in a discussion of Indian languages, Rafinesque casually made the most incredible claim: "Comparative Philology is now becoming in Europe the base of History and I have studyed it deeply, comparing 400 Eastern languages with about 85 American languages of which I have Vocabularies, and have succeeded to classify them."[60]

After seven letters from Rafinesque, exchanging views with Jefferson on ancient peoples, their languages and history, agriculture, and other matters, Jefferson terminated the correspondence by not even answering Rafinesque's questions about the professorship, pleading the infirmity of age and poor health.[61] John Patton Emmet of New York became professor of Natural History and other faculty were sought in Europe. In fact, there is no record of Jefferson presenting Rafinesque's name to the board. Earlier, Jefferson had rejected Rafinesque's application to be part of the Lewis and Clark expedition, and now for a second time he had thwarted his advances. Although Jefferson appreciated Rafinesque's erudition and brilliance, he must have realized—as Holley had discovered—that he would prove too troublesome. Jefferson had undoubtedly been warned about Rafinesque and was perhaps alarmed by the wildly imaginative forays that abounded in the papers Rafinesque had provided.

Rafinesque desperately pursued advertisements in newspapers and journals for academic positions. In answer to a request for a professor of language by the University of North Carolina at Chapel Hill, Rafinesque wrote a letter that dwelled upon his present woes in Kentucky, along with overblown claims of fluency in an astonishing array of languages. He was familiar with the literature of these languages, and he could write in them and speak them without accent, English among them. Unfortunately his letter abounded in ungrammatical peculiarities and mistakes, and despite an impressive list of references, Rafinesque did not get the job.[62]

Other employment opportunities fell under Rafinesque's gaze but no offers were made. Prospects were better at a projected Western College in Hopkinsville, Kentucky, one of whose members was his colleague and fellow botanist, Charles W. Short. Rafinesque suggested to Short that he be made president of the college. He also provided a plan for the organization of the college; fifty male and female students, each paying forty dollars tuition, would be admitted. Rafinesque would be paid $1,200 of the $2,000 collected while the remaining $800 would be used to hire two professors and pay all running expenses of the college. Short, a meticulous worker, was beholden to Rafinesque for his guidance and advice, but he did not approve of Rafinesque's sloppy science or his handling of plant specimens. Short remained civil and diplomatic, though his letters grew shorter and less frequent. Like Jefferson, Short must have dreaded receiving those nagging letters from Rafinesque.[63] The Trustees of the proposed college declined to consider this remarkably unrealistic plan and made no offer to its author despite Rafinesque's willingness to donate his one-thousand-volume library and his collections of plants and specimens. Rafinesque was rejected once again. The few positions available in the country's colleges and universities[64] would be filled by individuals less controversial and more reliable than Rafinesque.

Despite his seeming confidence and grandiose plans, an anxious Rafinesque was always aware of his precarious situation in America, and energetic by nature, he sought every possible means of support. Aware of the unstable and corrupt system of banking in the United States that resulted in the suffering of so many honest citizens, his thoughts turned to a new system of banking he had devised that would provide not only an income for himself but also would put an end to egregious banking practices for the greater benefit of society.

Chapter 8

THE WORLD OF FINANCE AND BANKING

> Wealth is power, Knowledge is power, Industry is power. Wealth should furnish the means, Knowledge unfold the ways, Industry effect them.

Rafinesque identified himself with the urban gentry of America, and though he enjoyed his association with the American middle class, he was never a part of it—this threadbare, spiritual aristocrat who hobnobbed with the rich. His colleagues accepted the fact that he was a kind of wizard, and being an intellectual ascetic, he was prepared to suffer lonely deprivation, living by his wits, and thereby achieving a certain independence. His aim was to spend as much time botanizing, collecting, thinking, and writing as was possible, but he was obliged to earn enough money to survive and to satisfy his compulsions. Collecting could be managed because he was not averse to living the life of a vagabond, and he gently imposed himself on fellow collectors and naturalists wherever he found himself.

Actually, Rafinesque was a confident entrepreneur who arose from a mercantile background, and he had a certain talent for making money in business. He seemed to have made a modest fortune in Sicily but had lost virtually everything in the shipwreck, a catastrophe from which he never recovered financially. He freely admitted (and complained) that as a natural historian and thinker who did not have a personal fortune, he lived a life of need and dependency. From time to time he was supported and advised by tolerant patrons such as John Clifford and Zaccheus Collins, both wealthy Philadelphia Quakers who had an interest in natural history. Unfortunately, both died relatively young, leaving their friend stranded without support.

In Rafinesque's scheme of things, it was the responsibility and the duty of the wealthy in a civilized society to support men like himself. Looking back,

Rafinesque mused wistfully: "With a greater fortune, or if I had not lost my estate several times by revolutions and shipwreck, I might have imitated HUMBOLDT, LINNEUS, PALLAS, KLAPROTH. . . . If Clifford had lived longer, he might have become for me the CLIFFORT of LINNEUS. But I have had friends and I have several yet: they know my zeal and steady efforts; they may yet encourage me or help me to pursue the laborious career I have traced to myself. Why should I not find protectors or enlightened patrons, as were found by Audubon and so many others?"[1]

While battered and vulnerable in Kentucky, Rafinesque applied his fertile brain to the problem of making money. Quite possibly, his involvement in the founding of a stock company that would finance the establishment and maintenance of a costly public garden had sparked an interest in money matters. Delving further, he devised the *Divitial Invention,* a banking scheme for his own and society's benefit. The invention, which he hoped would make him rich, was first announced to the world, without details, in the *Cincinnati Literary Gazette,* on February 26, 1825. A few months later he wrote to Collins:

> I have at last turned my attention to something practical and extensively useful, and have succeeded to achieve 4 Discoveries or Inventions of the utmost importance and Magnitude: Each of them is sufficient to change my State for the better; but I am going to apply myself to one after another in Succession and shall begin with the most valuable or profitable which I call the *Divitial* Invention, being a new Principle of wealth, which gives rise to a new Art, the *Divitial Art,* and a new banking System, calculated to cause a revolution for the better in Money Matters.
>
> I have applied and obtained a Patent for the same and I am going to Washington City to carry my Specification, which is so important that I would not trust it out of my hands. And from thence I mean to visit Baltimore, Philadelphia, Newyork, Albany and Boston, in order to spread my Discovery & put it into practice every where. I have also applied for Patents in England, France & other countries & mean to put it in operation at once in all those Countries. . . . I hope that you will not find my plans too gigantic.[2]

Rafinesque wrote Collins many letters, keeping him informed of his progress and bombarding him with his grandiose expectations. "If a new Divitial Bank can be established in every State, either by Charter, or by patent right, the Amount of Stocks for them, might be 50 millions, which at 1 per Cent only for patent right would be half a million! Then there is no end to this improvement." He confided that if his plan worked as well as he hoped, he might be able to "restore" the shaky currency of Kentucky.

Since the turn of the century, the newborn United States had been expanding its borders, acquiring new land by purchase, by seizure, and by military conquest until the freestanding nation, rid of European control, and increasing by one state each year (until 1821), reached from sea to sea. With the Monroe Doctrine of 1823, European powers had been placed on notice that their meddling in the affairs of the Western Hemisphere would not be tolerated. Settlers from the eastern states spread westward as the Republic began its explosive growth. There was an enormous need for capital to build newly founded towns and cities, as well as the canals, railways, and roads that connected them. Prospects of a limitless future prompted and justified massive borrowing and the founding of banks in every corner of the country. Money poured in from England. Fortunes were being made and lost in these financially unstable times when there was no convenient common currency in the nation, and the banking system and the stock market were largely unregulated.[3]

In a country teeming with aggressive entrepreneurs, sharp practice and fraud were not uncommon, bringing ruin to many innocent victims, Rafinesque included. Even some "respectable" financial institutions rigged the rules so that they made unconscionable profits, especially at the expense of the poor, a not uncommon banking practice that he despised. To him the bank should be a creator of wealth for the benefit of society, the honest laborer as well as the rich, and to this end he created the *Divitial Invention*, coupled with a system of Savings Banks to form *Divitial Institutions*. Just as assiduously as he inspected gardens, burial mounds, fossil sites, and the wilds, Rafinesque visited banks, especially savings banks, and thought about the improvement of their operation.

A patent for the Divitial system, obtained on August 23, 1825 by Rafinesque,[4] was advertised in the *Saturday Evening Post*. Any bank could adopt the system provided it paid a modest fee to the patentee. Later, Rafinesque established his own *Divitial Institution and 6 percent Savings Bank*, based upon the principle that the client could deposit any form of wealth in the bank—certificates of deposit (based on holdings of tobacco, cotton, wheat, flour, or rice in public stores), properly evaluated stocks, securities, banknotes, or specie. The bank would furnish the depositor with *Divitial Tokens* or certificates of equivalent value in various denominations—in effect, *money* of guaranteed worth that could circulate. Existing banks issued certificates of deposit, but these were not divisible, while in Rafinesque's system, divisible paper units or tokens were in effect, freely exchangeable units—a currency. Bearers of the tokens would be reimbursed upon their presentation at banks using the Divitial system. What really set Rafinesque's system apart was that deposits, which were converted to a currency, earned interest as bonds do; money would be earning money while tokens were in use. Dividends were to be paid to the customer on an annual basis, and the redeemed token would be worth more than when issued. Rafinesque used the term *divitial,* because according to him the word meant *leading to wealth.* Small investors investing their savings were welcomed, and banks and institutions were especially encouraged to participate in the plan.

Unfortunately, Rafinesque's scheme was launched in a sea of sharks. A dozen savings institutions mimicking Rafinesque's were founded by shady speculators (according to Rafinesque) at about the same time, in and around Philadelphia. These predatory institutions were secretly operated by the rich for their own benefit, offering less interest, making risky loans without adequate security (or no security if the borrower was a friend of the bank), and they were doomed to failure. When these banks failed, small investors lost their life savings while the owners of the bank lost nothing, because they had not invested their own money in the venture. Even worse was another kind of "deceptive institution"—the Loan Company was run by "usurious pawnbrokers," who paid depositors 6 percent on their money and loaned it at 36 percent interest per annum. Rafinesque had no complaint against the rich as long as they were honest and not greedy, but he railed against wealthy speculators who obtained loans at 6 percent interest without security while "the poor or industrious man cannot obtain a small loan at 6 percent *on a pledge.*" The Divitial Institution would make a loan of any size at 6 percent to rich or poor, but security was demanded of both.

Rafinesque's thoughts on banking and finances were summarized for a broad audience in a book he published in 1837, *Safe Banking including The Principles of Wealth,* a remarkable 138-page treatise whose frontispiece bears the ringing statement "*Every bank liable to risks or losses and calls is unsafe. Every bank liable to neither is safe.*" The book begins with an informative historical account of banking practices throughout the world from ancient times to the present. Lessons in economics and the principles of banking follow, with a detailed account of the corruption of financial institutions chartered by state governments, whose agents are eager to cooperate for favors received; these banks he labeled unregulated "engines of speculation." His voice was that of an economic realist, not a social reformer, nor did he ever bring politics into his discussion. Seldom does he speak of the social ills of the poor and the working man, although it was evident to him that laws often favored the rich and oppressed the poor. The enlightened Rafinesque preached to Americans that the wealthy elite must act responsibly in an egalitarian society—hardly a call to the barricades.

While only the government could issue currency in the form of silver and gold coins, banks were permitted to issue their own paper money, often without sufficient reserves to back it. They indulged in risky, speculative ventures, especially with land in the West at a time when American business and manufacturing were expanding at unprecedented rates. Even the generous law requiring one dollar of capital for every three dollars of bills issued was exceeded. In 1819, with the country afloat in paper money, an unstable credit situation resulted in a disastrous financial panic in which a large number of businesses and properties were acquired by the banks at a fraction of their worth as notes and credits were called in, and paper money was found to be almost worthless. Acquisitions were later sold by the large banks at immense profit after a return to their real value, to the detriment of smaller banks and investors. Rafinesque thundered against these unfair banking practices, and he warned that more than five hundred "insolvent" American banks were still in operation.

Though any bank or organization could use Rafinesque's system for a fee, Rafinesque eventually took part in the founding of a Divitial bank in Philadelphia, really a model bank that embodied the honest practice he advocated. By appealing to the public, Rafinesque managed to round up enough subscribers to acquire a capital of $50,000, and in June 1835 the bank opened for business, although not without "violent opposition by loan companies and unsafe banks."[5] The motto of the bank was "safety, utility, profit." It offered 6 percent annual interest on deposits for a specified length of time and promised to invest carefully and to have enough reserves on hand to satisfy any call. Promissory notes and unstable stocks were not accepted, and loans were not made without ample security; the important question of the inevitable fluctuation in the value of any stock was never clearly addressed. A complete record of the bank's investments and transactions was open to inspection by investors, and each year stockholders and depositors openly elected the trustees of the institution. In its first two years, the bank earned 17 percent for its clients. Still, the notion of the Divitial banking system did not spread, despite its apparent success and the endorsements of several respected individuals.[6] Rafinesque's earnings from his financial enterprise, which must have been considerable, are unknown, but were quickly spent publishing his writings, leaving little for his last years. Rafinesque's bank was in fact still in operation in 1840 at the time of his death, when he was in virtual poverty. Apparently he was cut out of the institution's operations, for he was not reelected a trustee of the bank, but reliable information about his involvement in the actual operation of the bank is lacking.

Despite the instability and the corrupt state of American banking, Rafinesque advocated a laissez-faire, open-market system that was free of government interference. He believed that "money, prices, values, rents, wages, and hire of money [would] regulate themselves and find their own level." The market would be corrected by reasonable and honest men responding to natural market forces. Rafinesque's discourses on economic matters reveal a lofty ethical standard and an idealism in his insistence that wealth be accessible to all. Although his writings on banking practices often sound didactic—in the style of a practical manual—at times his prose sparkles. *The Pleasures and Duties of Wealth* is prefaced with the genial:

> In deeds of good import your wealth employ,
> And happiness bestow, yourselves enjoy.

Though the use and duty of wealth is to "procure happiness to us and others," his experience induced him to write, "The besetting national sin of America is cupidity." In America, making money "fairly or unfairly is the great aim of life."[7] Much of his polemic sounds like a moralizing Sunday sermon as he speaks in generalities of tolerance and love, philanthropy, the good of the working man, and the value of labor, but the reader can have no doubt about whom he had in mind when he exhorts the possessors of wealth to "honor the men of Genius and Learning, publish or buy their works, their talents, the labors of their hands and minds, give them rewards and medals, make them happy in life and old age, do not allow them to fall in neglect to poverty." Obviously this was a cry from the heart, yet Rafinesque's transparent, intensely personal appeal was much broader; he exhorts his country to revere science and scholarship in order to be creative and prosperous.

To spread the good word about the new system of banking, it was necessary for Rafinesque to visit the financial centers of America. Upon obtaining official, unpaid leave from Transylvania University, he left Lexington in the spring of 1825, traveled by stagecoach through Ohio and West Virginia to the foothills of the Allegheny Mountains and then proceeded over this great barrier on foot, "as usual," into Virginia. His walks across the Alleghenies, which he repeated five times during his lifetime, seemed to take on the quality of a sacred, Romantic rite, which always harvested a crop of new plants. He was on his way to the great cities of the East, a pilgrimage to induce prominent citizens and banks to adopt his *Divitial System.*

He spent one month in Washington, where he met President John Quincy Adams and conferred with his old acquaintance from Lexington, Henry Clay, who was now Secretary of State. Clay, whose signature was required on patents to make them legal, was not sympathetic and was of little help in advancing his Divitial cause. His views mirrored those of a skeptical correspondent who asked Clay for his advice about the matter: "I should be pleased to know the probability of Mr. Rafinesque's success in his banking schemes, as he has flooded me with letters, appointing me *sole Agent* in all operations. I know him to be so visionary that I have given the subject but little attention."[8] Rafinesque also conferred with Richard Rush, Secretary of the Treasury, who was apparently interested, for he stirred up enough activity in the business and banking world for "agents protem" to be appointed and committees to be set up to study the Divitial system; Rafinesque was encouraged, reporting to Collins that there was talk of founding a new bank in Washington.

While in the capital, he kept busy. He took out patents for his *Divitial Invention* and several other inventions, and he began studies on winemaking and the cultivation of grapevines after visiting vineyards around Havre de Grace, Maryland.[9] Continuing his interest in the culture and languages of the Native American, he conferred with a Major McKinney of the "Indian Department" on the recording of vocabularies, with a plan to list at least one hundred words of each of the various American Indian tribes—a venture that he felt would interest Thomas Jefferson. He then journeyed to Baltimore and Philadelphia, conferred with bankers, visited gardens, and mingled with the local intelligentsia.

He was pleased to receive approval in Philadelphia for his Divitial Invention from Zaccheus Collins, a sober man with great experience in finance. Encouraged by the occasional kind word, Rafinesque had hoped to sell his Divitial Invention to the growing financial institutions in New York and Boston, but no sales took place, and when autumn had arrived, he was impelled to cut short his travels to return to Lexington. In October 1825, he wrote Collins that he was leaving the East with his mission unconcluded after spending more time and money than he had anticipated. He had only twenty dollars left to buy winter clothing and to travel six hundred miles to Lexington with books and luggage.[10] Collins lent Rafinesque thirty-five dollars, for which he was offered patents, Rafinesque's library, and "university emoluments" as collateral.[11]

Rafinesque's letters to Collins, with running commentaries of his progress, were boundless in their optimism, but characteristically in the same breath he was able to relate some devastating setback. He wrote: "I have now explained my Invention to 50 intelligent persons, and all think well of it; but the Banks are afraid that my Stock notes will be so good as to interfere with their Notes." Some individuals seemed to be interested in his plan, but not sufficiently convinced to pay royalties on the patent. Rafinesque complained that his "useful Divitial Invention was stolen or modified in Baltimore by establishing new Savings Banks partly on my plan, without consulting me nor asking my leave. A dozen have since been established; many are making money by it; while I the inventor who have spent $300 in travels, patents, advertisements, lectures, &c. to make it known have never realized a cent from it, for my expences and troubles!"[12] Though he had been "despoiled" of his invention, he never resorted to the courts to pursue his claims. "My dislike of law suits has compelled me to allow it till now."[13] As a fount of patentable discoveries, he learned that he must keep them "secret and in reserve . . . I was compelled to foil this kind of swindling or knavery by not taking any more patents."[14]

To spread the word about the Divitial Invention, Rafinesque informed people by letters he dispatched from a "Central Divitial Office" in Lexington on stationery whose letterhead proclaimed "Patent Divitial Invention." It would seem that out of nowhere, Joel Poinsett, the prominent American ambassador to Mexico, received a letter from Rafinesque, who was unknown to Poinsett, urging him to convince important Mexican officials to adopt the Divitial system. Rafinesque proposed that a bank be established in Mexico according to the principles of the Divitial Invention and that a company be set up whose purpose it would be to construct a canal across Mexico joining the Atlantic and Pacific Oceans. For this purpose, Rafinesque would help Poinsett obtain a loan from the American government. For good measure, Rafinesque added:

> As a further inducement for the Mexican Government you may acquaint them that I have made a dreadful Discovery in the Art of Defensive War. Or invented a *New Kind of Artillery*, a single discharge of which will destroy *One thousand Men and Arms,* one mile off, or sink a large Ship of War. This awful Invention will be communicated Secretly to all such governments who will grant me a Patent or Privilege for my *Divitial Invention.* I hereby authorize you to offer the knowledge & use of it to the Mexican government if they grant me the privilege asked above. . . . It will be sufficient to add that you may realize $100,000 yourself by helping me to establish my Invention in Mexico. Please to write me speedily, and send me a Duplicate of your Letters, under cover of the Secr of State Henry Clay my personal friend and fellow townsman.[15]

Poinsett never responded, even after a second, enticing letter. Apparently Rafinesque also offered these secrets to King George IV. Rafinesque's bizarre behavior would become known to all, enriching what was already gossip to become new legends associated with this fabulous character. On the whole, opinions were dismissive and intolerant, although some of Rafinesque's defenders would countenance no term that described their hero as worse than *eccentric* and insisted that he was the victim of a conspiracy created by people who were jealous of his great creative power. One can only conclude that as his Kentucky days drew to a close and he could not find a position anywhere, Rafinesque suffered serious mental derangement.

Chapter 9

TRAVELS AND FAREWELL TO LEXINGTON

His banking affairs in disarray, Rafinesque felt obliged to return to Lexington, burdened by his "unlucky detention" and failure in Washington and having not visited New York as he had planned. Finished with his financial and banking affairs for the present, Rafinesque left Philadelphia, traveled westward by coach through Pennsylvania to the foothills of the Alleghenies, and then for a fourth time he walked over the mountains to West Virginia and Ohio. On the way, he examined ancient monuments of Native Americans in Ohio, visiting museums and colleges where he eagerly sought invitations to lecture.

Rafinesque recorded what he saw in a series of notebooks, which he filled with descriptions, drawings and measurements, his first steps in the ordering of the world, and a permanent codification, understanding, and mastery of what he beheld in nature. He and other naturalists felt compelled to create a written record, a permanent account for future study; and although it gratified them and was indispensible, it dealt only with the surface of the wonders they beheld. Hoping to discern a purpose in their profound quest, a detailed description of nature was a training exercise, as they stuffed what they beheld into a bottle.[1]

A nasty shock awaited Rafinesque when he returned to Lexington; he writes that President Holley had "broke open my rooms, [had] given one to the students, and thrown all my effects, books and collections in a heap in the other. He had also deprived me of my situation as Librarian and my board in the College. I had to put up with all this to avoid beginning law suits." Outraged, Rafinesque gathered his belongings and took lodgings in Lexington, but he continued to lecture while preparing to depart. He gave a last course in botany in November 1825, and spent time in Frankfort, Kentucky, lobbying politicians and giving public lectures on the Divitial system at the behest of the State Legislature.[2]

Holley's behavior is surprising in light of his previous kindness to Rafinesque. Earlier, Rafinesque, a welcomed visitor to Holley's home, seemed to have a great affection for both him and his wife. While he had praised Holley earlier, he now cursed him and accused him of having a "hatred against science and discoveries." The abrupt change is perhaps less of a mystery in light of Holley's true interests and the hostile environment in which he labored. Holley was a classical scholar of religion, a New England Unitarian liberal, sufficiently enlightened to appreciate the value of science and natural history. He had been appointed president of the university over the strong opposition of a fundamentalist Presbyterian faction that never stopped agitating for strict religious instruction. In the ongoing battle between the liberals and the fundamentalists, he insisted that the university offer instruction in science and the professions and be open to all religious denominations, a policy that was defended by students, faculty, and the local press. But by 1827 the Presbyterian faction had garnered the support of the governor (and the state government) who cut funding to the university, an action they justified by the serious downturn of the state's economy. Holley, under great strain and harassed by jealous colleagues, resigned in protest in March 1827.[3] This thriving liberal institution headed by a popular president was badly damaged for many years thereafter. Enrollment dropped by more than half, by 1859 the medical school had transferred to Louisville, Kentucky, and the university veered toward becoming a theocracy, signaling the end of a period of Enlightenment.[4]

In this debilitating conflict among the many grievances of the orthodox, Holley had fought to permit Rafinesque to teach science, but his patience must have been tested to the limit by Rafinesque's frequent absences. As Rafinesque traveled and collected, he was rarely to be found at home base. He was forever making outrageous claims in the popular press and was embroiled in indefensible controversies—a noisy embarrassment to the university. Also troublesome was Rafinesque's aggressive national Divitial campaign, some of which could have bordered on the unethical. Why was a natural scientist spending his time promoting a bank? If he were preoccupied with banking and exploration, what time could he devote to teaching students and to caring for the library? One never knew where the unpredictable Rafinesque was, or when or where he would show up. However taxing it was for Holley to appease the relentless religious conservatives, a wandering, freethinking, theistic faculty member did his cause little good. By the time the peripatetic botanist and banker had returned from the East, Holley had had enough of both Rafinesque and the "bigoted religionists," and in a fit of pique, he broke into Rafinesque's rooms and manhandled his possessions, clearing space that was needed to house recently arrived, paying students. A decade later, Rafinesque wrote in his *Travels* that he had left cursing Holley and the university. History confirmed Rafinesque's special powers, for the curses "were both reached by them soon after, since he [Holley] died next year at sea of the Yellow fever, caught at New Orleans, having been driven from Lexington by public opinion: and the College has been burnt in 1828 with all its contents."[5] A somewhat overstated assessment.

Rafinesque wrote with some pride and inaccuracy that he was "never deprived of his Professorship and never resigned it." Almost penniless, Rafinesque required the assistance of a few sympathetic friends who created a fund to help him remove himself to Philadelphia. Some in Lexington considered him a visionary, a man of great learning but a failure in his ventures. For example, the great public garden he had struggled to establish languished under the feeble care of others. He departed from Lexington by stagecoach, "without regret," lecturing in Cincinnati and bidding farewell to his friends, about whom he admitted, "none was a CLIFFORD, who shared my taste and views."

Casting about for sanctuary during a troubled time, he considered becoming a member of the utopian New Harmony colony on the Wabash River, which he had visited on one of his earlier trips when it was controlled by George Rapp. Rapp had sold the commune to Robert Owen who, with William Maclure, was now running the enterprise. As Rafinesque became increasingly dissatisfied with Lexington and his desired academic positions eluded him, he gave much thought to this communistic community as a possible home, a safe haven.[6] New Harmony was organized according to a rational, enlightened plan, one that Rafinesque could not resist improving upon. At this very time, the colony was in full flower, although fatal flaws were becoming evident and would later put an end to the great social experiment. New Harmony could boast of a first-rate group of naturalists and geologists who had been brought from Philadelphia by William Maclure to initiate a superior grade of science in the Midwest. Still, a brutalized Rafinesque, now very cautious, felt that the colony did not really suit his needs, for it was too small and isolated a settlement, and too far from the East. Perhaps he was right.

Rafinesque's thinking about New Harmony was published in an article in *The New Harmony Gazette*, in which he outlined a plan that would provide societies such as this with an economic system under which they could thrive. What Rafinesque proposed was far more comprehensive than establishing isolated Divitial banks in the urban centers of America. In New Harmony, the

Divitial Invention could be a model economic arrangement for the benefit of an entire society—from "five to five thousand families"—living in a mutual association. In a Utopian scheme whose economic mainspring was the Divitial Invention, "no great capital is wanted, hands and tools are sufficient." The Divitial Bank coupon would be the convenient agent of economic intercourse, the lubricant. Money would no longer be the medium of exchange except with outsiders. Rather, "whatever has an exchange value, such as property, rents, wages, labor, &c may be used for mutual exchanges and cooperation at all times," using a divisible coupon that would represent the true value of the exchanged item, as determined by selected managers and appraisers. A commission of between 2 and 5 percent would be charged on all deposits, and money or stocks loaned to "strangers" (outsiders) would be charged 6 percent per anum. Profits, if there were any, would be divided among members or used to maintain libraries, schools, and other services. There would be no poor in an ideal system such as this. All in these "great families" would be housed, and "the rich would find a good and benevolent investment of their property."[7]

Rafinesque did not believe in depriving the rich of their wealth, but rather, he believed that providing a floor for the poor could eliminate economic conflict, so that no one would lack food, shelter, and clothing. In fact, Owen and others were trying to implement such a program that included the use of "labor notes," the equivalent of Rafinesque's coupons. But despite the efforts of the community, which operated under at least four revised constitutions in an attempt to become a stable state, great discontent gave rise to bitter words, and the commune collapsed after less than two years of operation. However the educational and research component supported by Maclure persisted and became a major factor in bringing science and progressive education to the near Midwest.

Maclure invited Rafinesque to join them, offering to pay transportation costs for his library and collections, but Rafinesque was wary of parting with everything he had accumulated since his shipwrecked arrival in America. In the end, he declined to join a group whose members he really did not know, where he was without a single close, protective friend. Indeed, relations with one member, Thomas Say, were strained, giving rise to Rafinesque's dark hint that some members of the colony were "jealous" of him. Isolated as he would be, promotion of his Divitial Banking scheme would be most difficult. He probably had enough good judgment to realize that this kind of arrangement might not suit him, individualist and urbanite that he was.

Cincinnati was his next big destination, where he lectured and then traveled on by stage toward Lake Erie observing and collecting. He traversed Ohio from south to north and paid a brief visit to another Owenite colony at Yellow Springs. He boarded the eastbound Detroit-Buffalo steamer upon reaching Lake Erie. All of creation was grist for his mill—including the fish of Lake Erie, the towns on the south shore (Cleveland, Erie), and the limestone formations in the area that bore fossils.

Niagara Falls, a mecca of the Romantic age and a must on any traveler's list, captivated Rafinesque as he observed the falls from both the American and Canadian sides. His response to the mighty falls was one of awe and unalloyed admiration, much in contrast to his feeling of "horror" upon viewing a belching Mount Etna, another wonder of nature. Pairing the two natural phenomena, he noted the contrast in the "sublime effects of water and fire."

Rafinesque continued his eastward journey, sailing along the Erie Canal, a recently built waterway designed to transport the goods and produce of the Great Lakes region and the Midwest to the Hudson River and the port of New York. A packet ship carried him from Lockport, in the Niagara area, to Rochester, where after learning of an expedition led by Amos Eaton he sought out the amiable professor of chemistry and experimental philosophy at Rensselaer School in Troy, New York. And according to Rafinesque, this was a school for teachers of science.[8] Eaton had organized an "experimental travelling summer school" in which he and his students leisurely cruised the Erie Canal on the *Lafayette,* stopping at points of interest for collecting "fossils, plants and minerals, and holding classes."[9] For the students, the adventure was memorable, the great success of the venture attesting to Eaton's knack of innovative teaching, not least of which was his training of women botanists.

Eaton wrote his wife: "When we were at Rochester, the celebrated Rafinesque overtook us. He joined our party and is now with us, and is to continue on to Troy." Rafinesque helped with the lecturing and tutoring. Joseph Henry, a young physics professor at the Albany Academy and later Secretary of the Smithsonian Institution, was also on this tour, and although aware of Rafinesque's flawed reputation, he learned from him and could not help admiring the man.[10]

In a setting such as this with students at his feet, Rafinesque was at his best, displaying his dazzling erudition and quick mind. Apparently he was endearing enough to his host and the entourage to merit an invitation to be a guest in Eaton's home while in Troy. Eaton recognized Rafinesque's peculiarities: "He is a curious Frenchman. I am much pleased with him; though he has many queer notions."[11] Because he and his students could learn from him, Eaton courted Rafinesque even as they found his eccentricity in dress and behavior amusing. In the midst of his troubles, lonely and friendless, Rafinesque was grateful for this admiring company and would always remember this canal trip with the greatest pleasure.

Although not the most creative of scientists, Eaton was an energetic teacher with broad scientific interests who vigorously promoted his discipline. About Rafinesque he said, "Even those who are disposed to pronounce Mr. R. an extravagant enthusiast, all agree, that he is a scholar of the first order; of vast reading and great classical learning. His nice discriminating talents have never been questioned." Eaton minimized Rafinesque's "supposed scientific heresies"—of confounding "shades of variety [of plants] as evidence of specific differences."[12]

Despite Rafinesque's patronizing and insulting critical review of Eaton's *Manual of Botany*, Eaton remained friendly, open, and supportive, and over the years they consulted each other about practical and scientific matters, each seeking the other's help and advice. Rafinesque even named a genus of grass *Eatonia* after his friend. Both Eaton and Rafinesque were able to forgive and forget insult and hard feelings with amazing ease. Thereafter, Rafinesque frequently returned to the Troy-Albany region where he examined the minerals and plants of the area and lectured to students an™d the public. In 1833, on one of his visits, Rafinesque and a companion, the Reverend Mr. Wiley, explored a "Bald Mountain," 1,200 feet above sea level, near Troy. The formation was called Mt. Rafinesque thereafter.[13] Such an act would suggest that Rafinesque's undoubted abilities were recognized and held in high regard. It would seem that in periods when he was not threatened, when he felt appreciated, Rafinesque behaved more rationally and could be truly impressive, meriting high praise.

After leaving Troy, he boarded a steamboat for a journey down the Hudson River to West Point where he conferred with another friend, the well-known botanist John Torrey, a professor at the Military Academy.[14] And then on to Germantown at the doorstep of Philadelphia, where there were several plant collectors who considered Rafinesque "one of the most eccentric, but one of the ablest writers upon America's natural History." It was from Germantown that a trusted, influential friend, Reuben Haines, drove him to a nearby commune at Valley Forge, the Friendly Society of Mutual Interests, which was again based on Robert Owen's socialist principles. After inspection and evaluation, he pronounced the community "disorganized" and fiscally unsound, and indeed its lifetime proved to be short. The society was disbanded in 1827, after bitter arguments about property rights and which moneymaking ventures they should establish. Perhaps a more compelling reason for not joining the commune was that Haines himself, a potential Rafinesque patron, declined to settle there. Haines, a wealthy dabbler in science (typical of the Philadelphia elite) and a prominent figure in the Academy of Natural Sciences of Philadelphia, was the proprietor of Wyck, a charming Germantown residence that welcomed natural historians and artists such as Thomas Say and Audubon.

Rafinesque resided at Wyck during August 1826, unable to proceed to Philadelphia because he did not have money enough to pay for rent in a boarding house. The correspondence at this time between Rafinesque, in Germantown (then outside the city limits of Philadelphia), and Zaccheus Collins, in the city proper, revealed Rafinesque's desperate situation and Collins' growing impatience with his friend. Rafinesque had made three attempts to see Collins without success. He needed advice and help because he was out of money, his baggage and apparatus were scattered, with some being lost, and he had nowhere to place his collections of books, plants, and minerals.

The bulk of his collections of plants and fossils, previously sent to Philadelphia, were impounded because he could not pay the cost of transportation. Unable to sell any specimens, it took him five years to reclaim his property with money that had come from the sale of his patent medicine Pulmel and his book *Medical Flora*. By this time, 10 percent of his specimens were ruined and the rest had deteriorated badly. Despite the offer of his herbarium of 36,000 plants, which he considered the finest in America, as an inducement, institutions declined to hire him. Rafinesque was bitter: "I should have wished to offer them to some liberal Institution that might have adopted me; but I have found none such in America as yet. They are yet, rich or poor, quite selfish like individuals, begging from all, seldom buying, never giving."[15]

Rafinesque considered becoming an itinerant lecturer: "I think I could collect handsome audiences in the states of N. York and New England for short Courses of 12 Lectures for $2 per ticket, on the new and popular subject of the Ancient and Modern history of America upon which I am prepared to *lecture extempore,* perhaps a Course could also be given in Philada." Worried about the safekeeping of his collections, their worth he unrealistically overestimated, he was, as always, optimistic about rescuing himself from this temporary financial embarrassment by lecturing, writing, and exploiting his Divitial system. He needed—indeed he begged for—a loan from Collins, his collections serving as collateral: "I think that $50 would set me up. . . . Please to help me if you can. I have not funds to advertise, else I could find some Employment. By sending me protem $15 (or even $5 at least) which would increase my small debt to you $50 . . . you would oblige me much."[16] Two days later Collins replied by sending him $15 and some stern advice. [17]

He advised Rafinesque to sell his collections in Europe where he could obtain a better price, and he discouraged the notion of his becoming an itinerant lecturer, for there was little money in it. In Rafinesque's letter, he had mentioned a grand plan for establishing an Owenite colony in Philadelphia. Perhaps such a community in an urban setting, with Rafinesque as leader, would provide him with security and a large, captive audience. Collins reminded him, coldly, of the sad fate of similar colonies. Why would this one succeed? Relations between the two men had cooled—Collins, exasperated and aloof, fended off an importuning Rafinesque, and there were no further written communications between the two men.

Chapter 10

THE MEDICINE MAN

Rafinesque was contemptuous of the medical practices of his day, resolute in his rejection of what trained, "allopathic" physicians had to offer, which in fact was very little. Most physicians were graduates of medical schools, but competency tests did not exist and licensing to practice was alarmingly inadequate. Many of their nostrums (antimony, calomel, opiates, arsenate, strychnine, prussic acid) and therapeutic procedures (bleeding, purging, blistering, heat treatment, leeching) were unquestionably useless. In an age without anesthesia and no understanding of sepsis, surgical operations were sentences of acute pain, intense suffering, or death. Rafinesque, when only nineteen years old, had wisely turned aside the great Dr. Benjamin Rush's kind offer of an apprenticeship, for he would have been inculcated with an outmoded medical philosophy that prescribed bleeding for many ailments. Rush had almost killed President Washington by exsanguination. When students began to question the master about the usefulness of this treatment and of administering massive doses of calomel, a semi-lethal mercurous chloride concoction, they were banished as traitors. While in Lexington, even when gravely ill with a virulent form of measles, Rafinesque refused the ministrations of a physician, spurning the "poisons" that had killed so many patients, and with Rafinesquian logic he proclaimed that his survival attested to the correctness of his view.

Most Americans, including President Jefferson, recognized the unsatisfactory state of medical practice. A "revolution" in medicine, not just reform, was needed, an opinion that was elaborated upon in a letter by the president to Dr. Caspar Wistar of Philadelphia. The physician "establishes for his guide some fanciful theory . . . of mechanical powers, of chemical agency, of stimuli, of irritability . . . or some other ingenious dream, which lets him into all nature's secrets at short hand. On the principles which he thus assumes . . . [he] extends his curative treatment. . . . I have lived myself to see the disciples of Hoffman, Boerhaave, Stahl, Cullen, and Brown succeed one another like the shifting figures of a magic lantern. . . . The patient, treated on the fashionable theory, sometimes gets well in spite of the medicine."[1]

In Rafinesque's day, the medical establishment was viewed as elitist and the preserve of the wealthy, and treatment was not readily available to all. If a physician was called upon, it was usually as a last, desperate attempt at healing. Most medical practice was, in fact, home practice by nonphysicians, caring for those on farms where most Americans lived. Unorthodox practitioners did not have expensive university training nor did they feel they needed it, for their success in healing did not depend on book learning. Families had their own nostrums, handed down from one generation to the next, and numerous books and pamphlets for the lay public on health and disease were very popular at a time when literacy was rapidly increasing, and technological advances had made printing and the production of books inexpensive and widely available.[2] The sale of self-help medical books and patent medicines for the public was a thriving industry then, as it is today. Between 1747 and 1856, John Wesley's *Primitive Physic,* a popular medical manual, had gone through thirty editions in England and America.[3] The public also had available the wisdom of numerous medical sects, each with its own cures and philosophy of healing, and all bordering on the fraudulent.[4] Ignorant quacks abounded, each pushing his own moneymaking patent medicine. One such was the father of John D. Rockefeller Sr., a confidence man who went about the country peddling a cure for most diseases, including cancer. The patent label, sanctified by governmental authority, meant little because the effectiveness of a medicine did not have to be demonstrated.

Rafinesque, pragmatic as always, was interested in the application of botanical knowledge to the treatment of disease, and he was never uncomfortable in the role of a botanical doctor or a manufacturer of medicines, botanical or mineral. Although he had no degree, he lectured on medical botany at the Transylvania Medical School, and as a naturalist roving over many states, he conscientiously sought plants of potential therapeutic value to the physician. Plant materials were used not only by "unqualified" physicians but also by trained physicians such as Benjamin Rush and G.B. Wood, and in time the practice of herbal medicine was standardized in the publication *Plant Materia Medica*, an inventory of plants, their descriptions, and uses.

Along this line, Rafinesque wrote an affordable, portable botanical manual in two volumes (1828, 1830), the *Medical Flora, or Manual of the Medical Botany of the United States of North America,* which he believed provided thoughtful physicians, students, and pharmacists with useful information about medical botany and helpful suggestions, which would be of value in the home. This proved to be Rafinesque's most profitable book.

Medical Flora was systematic and informative, with helpful glossaries of botanical, chemical, and medical principles, as well as medical properties and chemical tables—a kind of precursor to the modern *Merck Index.* More than any other publication at the time, *Medical Flora* was a treasure house of information on botany applied to medicine, which Rafinesque had gathered from many sources throughout his entire career—Native Americans, colleagues, and books—traveling 8,000 miles through fourteen states, mostly on foot, close to his botanical prey. Rafinesque freely credited many individuals for the information they provided, and he stated that he was perfectly willing to listen to the learned, as well as the illiterate. He never "despised knowledge because [it was] imparted by an uncouth mouth."

However, many regular physicians were hostile to an author who had the effrontery to pass himself off as a medical doctor—indeed, a lung specialist. Understandably, he was detested by the authors of virtually all competing botanical works that he had judged harshly and summarily dismissed. Of Barton's *Vegetable Materia Medica* he wrote: "Another costly work mentioning about 1 plant in 40 of North America. Descriptions short and flimsy"; Cutler's *Plants of New England:* "Rude attempt, many botanical mistakes, some medical indications"; Henry's *Medical Herbal:* "Empirical, erroneous in names, descriptions, facts and figures, some medical facts, and local names"; Hunter's *Narrative:* "Another impostor, he has given a list of western medical plants with Osage names, not to be depended upon or ascertained."[5] Using words and phrases such as "imperfect," "worthless," "a mere herbalist," and "a spurious Botanist," Rafinesque earned himself new sets of enemies.

The major alternative to conventional American medicine in the nineteenth century was *Thomsonism,* founded by Samuel Thomson, a New Hampshire farmer who established a system of practice based on "a course of six numbered remedies to be given in sequence."[6] Thomson's promotion of his brand of medicine was ingenious and effective, going beyond the use of specific medicines. His practice was in a sense holistic, incorporating elements of social manipulation used by Utopean communities like the Rapp Colony, New Harmony, and Brook Farm, where subscribers came to feel that everyone would be cared for—an assurance that in itself was a critical part of therapy. In 1811 he established the *Thomsonian Friendly Botanical Society,* in which clusters of families dedicated to mutual aid and the use of his remedies could join the society for twenty dollars. Purchasers received a certificate outlining their rights and privileges, along with Thomson's book, *Family Botanic Medicine,* which included a short course in botany and a list of therapeutically useful plants and their applications, case studies, and numerous testimonials from patients who had been given up for dead. His writing, a prolonged defense against denunciations by physicians, was embellished with cracker barrel philosophy, scripture, poetry, and numerous references to the ancient Greeks. Thomsonism flourished in a land where it was believed that everyone could be his own physician. His medical procedures were simple and easily understood by all, so that physicians were really not needed. At one time he faced a charge of willful murder instigated by a physician, but this persecution only inspired him to pursue his holy mission with greater zeal, fending off enemies, and expounding on his medical philosophy that "all diseases are the effect of one general cause and may be removed by one general remedy." Thomson introduced a Jacksonian element of social and economic conflict in his program—the common man against an entrenched elite. His aim was to provide inexpensive medical care for the poor that was not offered by the university-trained medical establishment which was drilled in subjects such as human anatomy, which he deemed valueless.

Thomson hired agents to establish societies and sell both his books and botanical preparations. With characteristic Yankee energy and aggressiveness, he became enormously popular and for a few years constituted a real threat to the medical establishment. There were, however, several other competing systems of medical practice at the time. These other systems, compounded with dissension in the organization, resulted in a decline in the success of Thomsonism. By 1840 the self-help movement had languished and the American Medical Association had been founded. Conventional medicine began to improve as standards were established, and modern medical practices from Paris were imported by a young generation of American doctors.

Rafinesque never embroiled himself in the polemics of internecine warfare between different medical groups, just as he would not involve himself in American partisan politics. He could not wholeheartedly subscribe to the tenets of one or the other of the numerous medical cliques. Serenely aloof, he was an active practitioner of his own brand of medicine. Disease to Rafinesque was a problem to be solved, each case a research project to which he would apply his power of reasoning and his impressive store of knowledge.

Rafinesque separated medical men into three general categories.[7] The first group, whom he admired and with whom he loosely identified, *Rationals,* were "liberal and modest, learned, and well informed, neither intolerant nor deceitful, and ready to impart information." *Rationals* were of three types—*improvers*, who studied nature and enhanced knowledge, *experimentalists,* who "are directed by experience and experiments, observations, dissections and facts." The third type was the *eclectic,* who adopted what was found to be most beneficial for the patient and was "willing to change according to acquired knowledge, circumstances and emergencies." Apparently, this was the first time the word "eclectic" was ever applied to a school or class of practitioners whose thinking was not dominated (and crippled) by adherence to an all-encompassing medical theory or to a particular regimen or set of drugs. The eclectic school, also called the *Reformed Practice of Medicine,* prospered, establishing schools and starting journals. Rafinesque was credited with founding this prominent movement, but in fact, later eclectic physicians had only borrowed Rafinesque's label and little else.[8] In his last years, Rafinesque admitted that he was closer to the "eclectic" school in spirit than to Thomsonians, homeopaths, and botanical empiricists, all of whom he despised.[9]

His harshest words were saved for the *Theorists,* whom he berated as "illiberal, intolerant, proud and conceited," who segregated themselves into secretive sects (Galenists, Brownists, Mesmerians, Calomelists) at conflict with the world. *Empirics,* consisting of standard, practicing physicians, herbalists, Indian root doctors, steam doctors, and quacks, he described as "commonly illiterate, ignorant, deceitful, and [who] follow a secret and absurd mode of practice, or deal in patent remedies." The latter group was in desperate need of "instruction in the natural knowledge of medical substances."

Soon after Rafinesque returned to Philadelphia in 1826, he contracted what he diagnosed as "catarrhal and dyspeptic consumption," also called by him "fatal phthisis," brought on by "disappointments, fatigues, and the unsteady climate." Having no faith in conventional therapy for tuberculosis, which indeed was of little or no value, he began treating himself with "a combination of several powerful vegetable substances" of his own devising, and miraculously, he restored himself to health. Whether he actually suffered from tuberculosis or consulted a physician is not known. At the time, physicians were very familiar with the disease, so that diagnosis may very well have been correct. If indeed it was tuberculosis, the disease sometimes arrested spontaneously and was not invariably fatal. Nevertheless, he credited his cure to a patent medicine of his own devising, *Pulmel*, a discovery that he sincerely believed was a cure that should not be denied to the world. Rafinesque announced the breakthrough in an article entitled *Pulmel, Great Medical Discovery,* signed *Medicus* in the *Saturday Evening Post* of June 16, 1827. He used an alias because he did not want his name associated with the controversy and the inevitable censure that would follow from his claims, but the secret did not last long. Not only was the nostrum advertised as a cure, it also *prevented* the disease. Pulmel, which had a pleasant taste and smell, came in a variety of forms—in a chocolate drink, a syrup, a lozenge, a lotion, and in a powder or pills. If the disease affected the lungs, a perfume of Pulmel (a *Balsam*) could be inhaled. Rafinesque arranged for the manufacture and sale of Pulmel under his supervision in its various forms with several Philadelphia pharmacists.

On several occasions he had written about the adulteration of known medicines by inferior or even mislabeled ingredients, but for Pulmel, Rafinesque obtained the finest herbal materials from a shaker colony in New Lebanon, New York, a source he recommended to all. Shakers issued catalogs and became the major suppliers to doctors and pharmaceutical companies in the nineteenth century. The flourishing enterprise was nourished by Dr. G. K. Lawrence, a Shaker physician and medical botanist.[10] Indeed, information furnished by Dr. Lawrence found its way, verbatim and with full credit, into Rafinesque's first volume of *Medical Flora*.

Rafinesque would not divulge the ingredients of Pulmel other than to state that they were of botanical origin and that the active chemical was *Pulmelin*, a salt of Pulmel.[11] An impoverished Rafinesque felt that on several occasions he had been cheated of income that was rightfully his: "I ought to have made another fortune by my Inventions, which comprised so many things. . . . I was compelled to foil this kind of swindling or knavery by not taking any more patents." But this was not going to happen again: "Happily I had kept my secret [Pulmel], and in reserve some other discoveries of mine." The composition of Pulmel has never been revealed.

He made sufficient profit from the sale of Pulmel to publish some of his work—in 1832 he recorded that he made $263.87 and spent $190.72 on publishing, $92.80 on travel, and $98.15 on food. This unhappy arithmetic obliged him to withdraw funds from his meager savings. He published several articles on the efficacy of Pulmel in the *Saturday Evening Post*, a journal that also carried advertisements for his remedy. Ready access to the journal becomes less surprising upon learning that the editor was involved in the distribution of the elixir. Rafinesque's optimistic reports, one entitled "Cheering news for those who have consumption," briefly summarized progress he had made in treating patients—anecdotal evidence of cures[12]—and in 1829 he published a seventy-two page booklet entitled *The Pulmist; or Introduction to the art of Curing and Preventing the Consumption or Chronic Phthisis,* a fuller discussion of the subject. His justification for writing the book was that he had restored himself to "perfect health and a sound constitution," and had "met with repeated success in my experiments and practice."

Rafinesque, ever the entrepreneur, not only invented, produced, and promoted his "radical cure," but he also became a medical practitioner, a *pulmist,* specializing in diseases of the lung despite his lack of formal credentials, namely a medical degree. On the title page of the book he added the *Ph.D.* degree, and *Pulmist* after his name, and listed the numerous positions he held and honors he had received: "Professor of Practical and Medical Botany, Natural History, &c &c, author of the manual of *Medical Botany of the United States, the Analysis of Nature,* and 50 other works or pamphlets. Member of the Medical Societies of Cincinnati and Lexington; the Philadelphia Society and Lyceum of New York; the Academy of Natural Sciences of Philadelphia; the American Antiquity Society of Worcester and Nashville; the Kentucky Institute, &c; and of several learned Societies of Europe, in Paris, Bruxelles, Vienna, Bonn, Florence, Naples, &c."

Paragraphs in the book were numbered as in the Bible. He began reasonably by stating that he did not want to interfere with the working of the family physician. He desired "the cooperation of all liberal physicians" rather than being "deemed an opponent" though he admitted, "I have often found to my sorrow that their practice is so erroneous and incompatible with mine."

There followed a discussion of the history of tuberculosis, the incidence of disease with the occupation of the patient, and the spread of this "hereditary disease" by "cohabitation, contact of the breath, and putrid expectoration." Consideration of the influence of temperament on susceptibility—of bilious, sanguine, melancholic, and phlegmatic personalities—echoed the past. Rafinesque, always the species splitter, invented new categories of personality and gave them remarkable names—athletic, muscular, nervous, serofulous, dermic, and leucodermic. Sixty forms of tuberculosis were created, named, and discussed, and while some discussion was insightful, there was much twaddle. A form of tuberculosis that he called *Moral Phthisis* was caused by "passion, love, nostalgy, disappointment, exalted susceptibility, &c." It was curable, but "difficult when connected with a broken heart." Rafinesque seemed to be entering the realm of psychosomatic medicine, and his awareness of an association of disease with occupation and emotion was remarkable.

He tried Pulmel on "perhaps one thousand individuals," and of the 120 cases followed, he claimed that sixty received some benefit and twenty-five were cured. Another discussion of tuberculosis, which he called "the plague of northern climates," was filled with interesting, if unbelievable statistics. Though one sixth of Philadelphians died of the disease—and once contracted only one in twenty survived—by employing Pulmel, Rafinesque found that ten in twenty were saved and that the other ten received some relief!

Rafinesque as a physician did not have a conventional office practice. Rather, he corresponded with his patients, who learned of him and his cure through advertisements—which advised them to purchase Pulmel from certain Philadelphia pharmacists. If prospective patients would furnish him with details of their illness and enclose $10 he would send them by post "a full consultation in writing." Assuredly, Rafinesque believed he had discovered a cure for a terrible disease and was practicing medicine for the good of humanity, but there is little question that the desperate plight of humanity afforded him a much-needed income. From one of his patients he received $50, which quickly disappeared by paying off various debts, enabling him to recommence an active publishing career.

One such consultation, four pages in length with a cover letter in Rafinesque's hand, has survived.[13] Despite Rafinesque's odd theoretical pronouncements and the aura of half-baked chicanery that surrounded him, one cannot help but be impressed by the sober, cautious, and enlightened advice given to his patients. However, there was nothing truly original in his treatment, for not only were many physicians advocates of the same clinical management, similar medical advice could be found in popular books on home management of disease and in almanacs, many published in Philadelphia.

When dealing with the suffering of ill people, Rafinesque's opinions became reasonable and enlightened, and his intelligence shone through. Few physicians could have done better at that time, and many did worse. His first rule, well-known but often ignored, was to do no harm to the patient who should not be made to suffer from heroic treatments. After analyzing the details submitted by one patient, he proclaimed his condition "quite curable with proper remedies, diet and Care," and deplored the fact that the patient had been previously treated with calomel and by bleeding, which made his condition worse. Fresh air, some exercise (on horseback), and the wearing of warm clothing were beneficial. Some of his advice, while foolish by present standards was at least not dangerous or harmful; he recommended the wearing of a piece of rabbit fur, "the hair inside, suspended by a ribbon from the neck" over the area of pain. This he felt was "almost a rubifacient in effect, keeping the breast electrified, warm, comfortable, and easy, being a complete substitute for a blister," without attendant irritation or pain. In addition to forty doses of Pulmel over time, he recommended several other mild tonics, sedatives, and remedies for specific complaints—constipation, diarrhea, and coughing—and certain drugs, none "pernicious poisons," to be left up to the patient to decide whether they were truly required. Surprisingly, he did not harp on the benefits of Pulmel in the consultation.

Diet was an important component of his treatment. He disparaged a starvation diet (recommended by some), or "gross, raw, tough, salted smoked & fat meats." He urged the patient to eat rich broths of meat, turtle soup, a variety of greens, fruits and vegetables, oysters, rice, "Eggs in any shape . . . tender well done meats, chiefly Fowls, Pidgeens, Lamb & Veal." He advocated the drinking of milk, and regarded tea and coffee as "useless."

The consultation was a respectful, rational dialogue with a patient whom he encouraged to modify several specific but undogmatic suggestions as required, and he invited responses from patients. He encouraged patients to follow their conditions by determining pulse rate and temperature. Rafinesque seemed so reasonable and grave when he wrote: "You must of course use moderation in everything. You are a married man & you will understand me. I do not prescribe Privation but moderation in everything, Eating, Drinking, Talking, Sleeping, &c."

Although Rafinesque was called a quack, he was different in that he passionately believed in the efficacy of his treatment. He may have misled himself and others, but he did not lie. He had been denied access to medical school and formal training, but in truth he realized that this was where the future of medicine and therapy lay. In his *Medical Flora* he states that medical science was improving because it was "borrowing" from all the natural sciences, and that pharmacy was becoming a science with the help of botany and chemistry. He knew what science was and was as cognizant of its potential as were the best physicians and academics.

Not only were the unlicensed a threat to the physician's livelihood, many were so ignorant and misguided that they sometimes killed their patients. In an ongoing battle, the medical establishment vigorously denounced the intrusion of ignorant quacks, tracking their publications and their activities and attacking them with broadsides. Dr. Daniel Drake, a professor of medicine at Transylvania University and editor of *the Western Journal of the Medical and Physical Sciences*, was one of the leaders in the battle. In 1830 he reviewed in his journal several books by nonestablishment medical people, which included Samuel Thomson, the inventor of Swain's Panacea William Swain, and Rafinesque, who had published his book on Pulmel. Drake had nothing but contempt for these men.[14]

He denounced these "unblushing authors" whom he ironically addressed as *doctor,* as moneygrubbing impostors with charms and incantations, and "barefaced and abominable quackeries." His extended polemic fulminated with mockery and malevolence as he pointed out the numerous errors and "the thousand and one fooleries" in their texts. Yet he and his colleagues had little more to offer than did Rafinesque, and a fair reading of Rafinesque would reveal no more error or "quackery" than in conventional medical texts—but Rafinesque was a marked man. Drake recounted the basis for Rafinesque's questionable reputation as a historian and the fakery of his Pulmel, suggesting that he print on his labels:

> Ye *would* be dupes and victims, and ye *are.*

Rafinesque's medical career was short-lived, a temporary enthusiasm lasting only about four years, and by the early 1830s, advertisements and articles extolling the virtues of Pulmel had ceased to appear. He was once again involved with his first love, natural history and science, as well as ethnography and a scattering of other subjects—too busy to spend time writing lengthy, detailed consultations. This source of income dried up, while the irresistible pursuit of nature condemned Rafinesque to a life of penury.

Chapter 11

HISTORY, ARCHAEOLOGY, AND LINGUISTICS

Rafinesque was rooted in the age of the Enlightenment, with its deist view of the world. Earnest and thoughtful investigations of human origins and history were not uncommon as the biblical version of human beginnings was increasingly found wanting and mythic. Radical thinkers were formulating the idea of a world based on reason, morality, and universal physical laws, presided over by a God of limited jurisdiction—a chief architect and little more. According to Genesis, all humans arose from a single act of creation, and the present-day assortment of human "races" was the result of change induced by exposure to different environmental conditions (food, climate, temperature, and myriad other factors) over long periods of time. This *monogenic* theory, accepted by the church, was held by major European scholars such as Blumenbach and Buffon, but was rejected by others, such as Voltaire and Louis Agassiz, who proposed a *polygenic* theory in which each of the human "races" was created separately. The issue was regarded as a conflict between rational science and orthodox church doctrine. Such gravely considered questions ultimately fueled interest in archaeology, anthropology (physical and cultural), and the comparative study of languages in an attempt to define the origin and limits of human variety and to establish the relationship between human groups in history.

Shadowing the religious and intellectual concerns of human genesis was the question of the origins of white masters and black slaves, a relationship that Rafinesque felt must have evolved within human groups since the Flood. If they were of a common stock, they were blood brothers, created as "equals" as the Constitution read; supporters of slavery could find little comfort here. On the other hand, if they were created separately and were of separate species, whites would have less difficulty consigning them to an inferior status and enslaving them as beasts of burden. Both camps strove to find evidence to support their positions, giving rise to numerous learned tracts and broadsides supporting one or another theory of human origins. Rafinesque firmly believed in the single origin of humankind. Although the issue was not a dominating concern of the day, it was not insignificant, and the only hope of resolving the issue was to acquire more data, a task taken up energetically by Americans in a land where slavery flourished. Much as he abhorred slavery, Rafinesque took no part in abolitionist movements.

In Rafinesque's America there was still another cause for interest in the problem of human origins and history. Whites were interacting with and displacing Native Americans from their lands. The origin of these people was obscure, their languages were strange and unknown, and their history only hinted at by mounds and earthworks, signaling a mysterious antiquity. Europeans were intensely curious about these people, either idealizing them or fearing and denouncing them as barbarian. Some explained their existence within a biblical context (one of the ten lost tribes of Israel). But to expansive entrepreneurial Americans, they were underfoot—in the way of national conquest and fulfillment. According to some, civilized people who knew what to do with land had every right to displace the ignorant native who, seemingly, did not use the land properly. Scripture was used by them as a self-serving justification for the complete dominion of humans over animals, and this notion was easily extended to encompass the inferior, "uncivilized savage."[1] The egregious treatment of these populations at the hands of the Spanish as they overran Central and South America and Mexico was now accorded to the natives of the North, as white Americans who had consolidated eastern lands trekked westward. For the white man's purpose it would be best to create a history for Native Americans that would dehumanize them and reduce them to a lower order.

Rafinesque's sympathies lay with the Native American. Commenting on William Robertson's *History of America*, he dismissed it as "a slanderous sketch of the native American," and in a published letter he expressed annoyance with those who spoke of the *Red man* of America with its pejorative undertone: "there is not a Red man (nor ever was) in this continent. There are humans of many colors, but none red, unless painted!" Rafinesque took a proprietary interest in America and its original inhabitants, defending a transcendent greatness that was not always appreciated: "The continent of America has ever been the field of philosophic delusions, as Africa of fables and monsters, and Asia of religious creeds. All the various systems of monks and philosophers on the origin, climate, inhabitants, language, &c. of America have been repeatedly destroyed by facts."[2] Rafinesque argued that the number of domesticated animals possessed by a people was a standard of their civilization, and drawing up lists of these animals—quadrupeds, birds, reptiles, fish, insects, and even worms—he showed that they totaled 112 in America, while there were only eighty domesticated species in Asia, Europe, Africa, and Polynesia combined. The paper abounded in arcane information about pets and other animals, which he gleaned from numerous sources that had been published over the centuries.

Rafinesque wrote about William Penn's 1685 treaty with the Indians: "It was the first instance of a colonist having bought a country from an European king, who had no more right to it than the king of the moon, buying again from the real owners of it. . . . Yet the good W. Penn did not pay the full value to the 10 ignorant Indian Chiefs, and his example has been closely followed to this day." With obvious disapproval, he recounted how Penn bought the land worth ten dollars per square mile at the time for fifteen cents per square mile, using as currency all sorts of baubles—beads, mouth harps, and looking glasses—tools, guns and powder, as well as beer, "strong water," and tobacco.[3] Rafinesque, by abhorring the cruel treatment of the Native American, antagonized influential Pennsylvanians for he had ridiculed the revered William Penn and had repudiated the sacred legend immortalized by the painter Benjamin West (Penn's Treaty with the Delaware Indians at Sackamaxon).

Refusing to categorize humankind narrowly, he asserted that a single act had created humans in whom were expressed a vast range of characteristics—sizes, body and facial types, and twenty shades of color (all but red, blue, and green). The intermingling of the various nations was prevalent in their long history. Rafinesque was expressing his opinions (consistent with those of Quakers) at a time when the scientific basis of racism was being established by Samuel George Morton of Philadelphia, an official of the Academy of Natural Sciences. Morton's measurement of cranial volume (brain size) led him to the conclusion that the volumes of black people were less than those of whites,[4] an erroneous conclusion that supported the "scientific" argument that black people had appeared on earth by a separate act of creation, and that they were of lower intelligence, and were, therefore, inferior.

Rafinesque's pronouncements were indistinguishable from those of the enlightened, modern commentator: "Let us pause before we form opinions out of a few facts. Truth can only be detected by extensive observations. Respecting mankind the result of those made all over the world demonstrate that man is a variable being, like every other, and subject to the ETERNAL DIVINE LAW OF PERPETUAL CHANGE AND MUTATION in form size and complexions as well as manners and improvements. Whence we ought to love each other whatever be our shape, bulk and hue, as brothers of a single great family."

Rafinesque believed that the differences in human groups were in relation to their environments. "The white man became tawny by constant exposure, brown in warm climates, coppery in cold regions, and black in the sands of India and Africa. The Mongol features had origin in the deserts of Northern Asia, and the negro features in those of Southern Asia and central Africa." Emphasizing variability, he goes on: "There are Mongols with different complexions, white, pale tawny, yellow, olive, coppery, &c; and there are white, yellow, brown and black negroes."[5] Rafinesque deplored the widely held notion that all early Americans were of a coppery tone. Rather, he emphasized the diversity of "complexions," pointing out in his overblown manner that from his reading he had identified at least twenty white American tribes—Parians, Guaijacas, Barvas, Chiquitos, and others.[6] Deploring divisions of mankind on the basis of color, he felt that if color is to be used, the terms *pale, tawny,* and *dark* should be used.[7]

In two publications, Rafinesque dignified the black people by constructing a sympathetic history for them, an account of their origin and their legends, comparable to those of other nations. In his *The American Nations,* published the year he died, he wrote, "Nobody has undertaken, as yet the history of the Negro nations: a labor so difficult and luckless as to be despised. My memoirs on this despised race, may perhaps furnish the basis of such a history."[8] One of his astonishing memoirs dealing with the "Asiatic Negro" showed that their language was related to those of the "African and Polynesian Negroes, as well as with the Hindus and Chinese, and rendered it probable that all the Negroes originated in the Southern Slopes of the Imalaya Mountains" just as other humans did.

The second remarkable paper was on Negroes in America *before* Columbus—*Memoire sur l'Origine des Nations Nègres*—in which Rafinesque identified and described twelve black tribes in America, all with related languages. The essay was sent to Georges Cuvier of the Société de Geographie de Paris, which had offered a prize for the best essay on the subject, and Rafinesque won the gold medal for his entry, one of the few signs of recognition Rafinesque ever received, that attempts to link "the languages and traditions of all the nations of the world with the primitive cradle of mankind" by historical, ethnographic, and philological analysis. The work was never known by his fellow Americans, and the prestigious award he received from one of the great societies of the world, perhaps the first by an American, was all but ignored, prompt-

Portrait of Rafinesque as a young man, probably by Mathew Jouett. Transylvania University Library, Lexington, Kentucky.

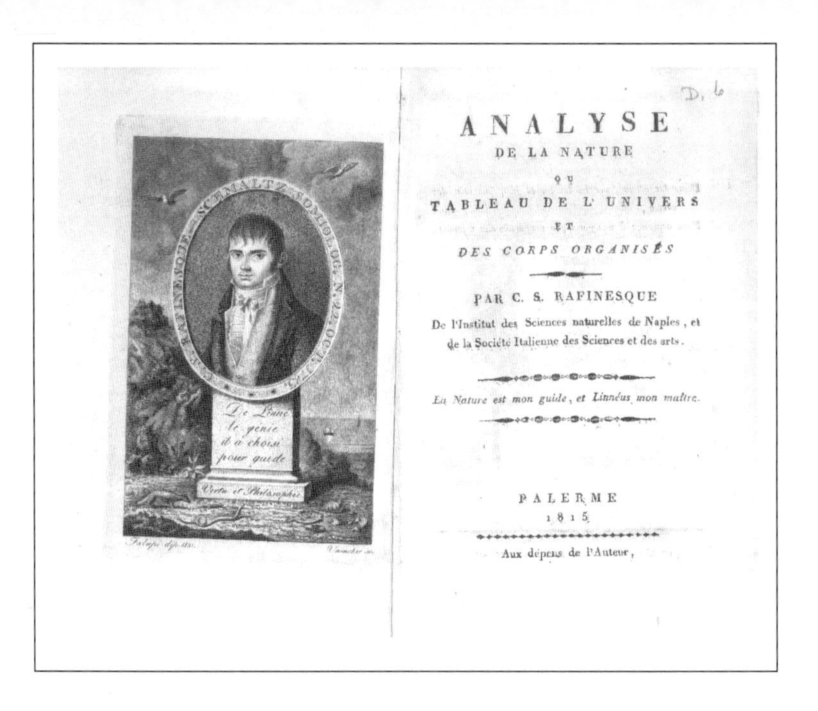

ANALYSE
DE LA NATURE
OU
TABLEAU DE L' UNIVERS
ET
DES CORPS ORGANISÉS

PAR C. S. RAFINESQUE
De l'Institut des Sciences naturelles de Naples, et de la Société Italienne des Sciences et des arts.

La Nature est mon guide, et Linnéus mon maître.

PALERME
1815

Aux dépens de l'Auteur,

Title page of *Analyse de la Nature* (1815) and frontespiece bearing Rafinesque's portrait by Falopi, dating from 1810. The inscription on the pedestal attests to his admiration for Linnaeus. The many forms of animals and plants of the earth and the sky reminds the viewer of his broad interests (Library Company of Philadelphia).

The main building of Transylvania University in 1818. Engraving from a drawing by Mathew Jouett, from Caldwell, 1828. (Annenberg Rare Book and Manuscript Library, Van Pelt-Dietrich Library Center, University of Pennsylvania.)

Sketches of a woman (left) and of Rafinesque's mother (right) by her son. These and more than twenty other sketches are owned by Transylvania University.

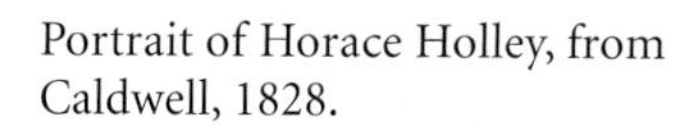

Portrait of Horace Holley, from Caldwell, 1828.

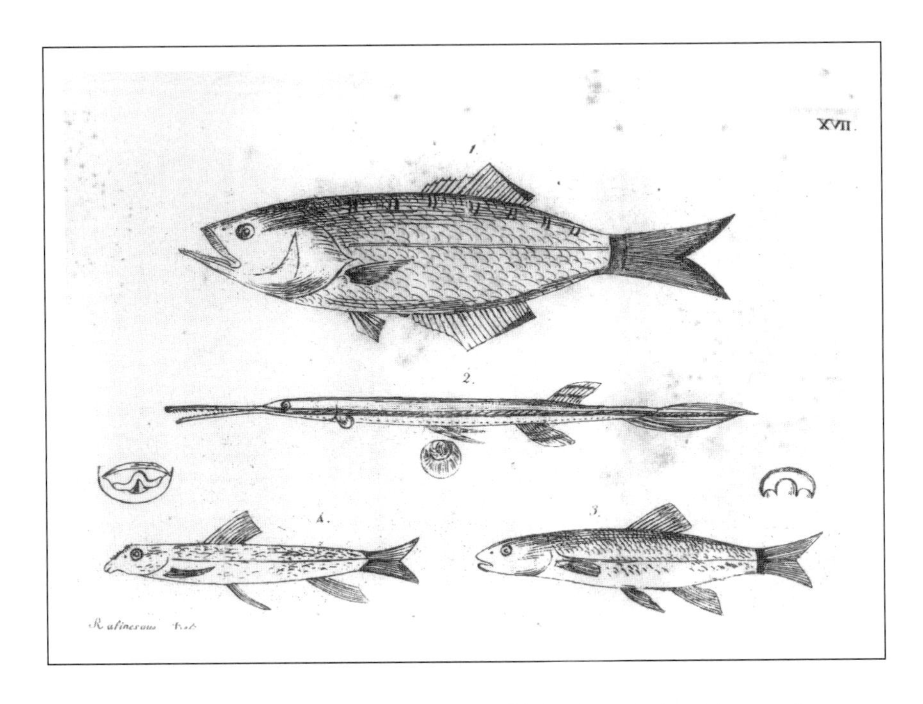

Rafinesque's drawings of three "new" genera of fish, *Pomoxis, Sarchirus, and Exoglossum*. From *Journal of the Academy of Natural Sciences of Philadelphia,* vol. 1, part 2, 1818.

Gold Medal awarded to Rafinesque by the Société de Géographie de Paris for his essay on the origins of the Asiatic Negro, 1832. Medal is 32 x 2.5 mm in size. (College of Physicians of Philadelphia).

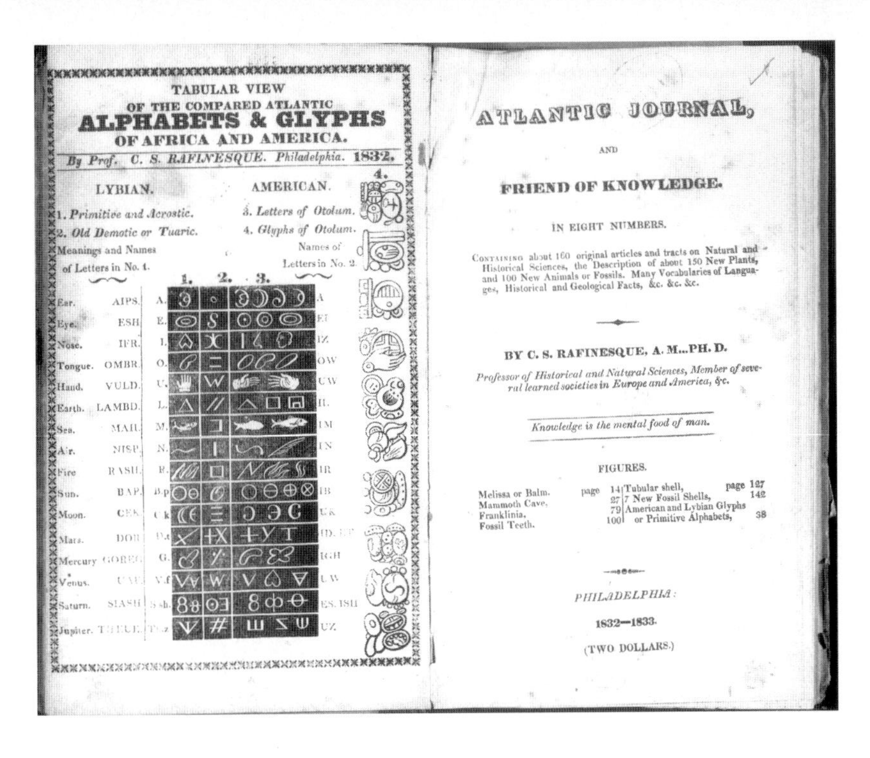

TABULAR VIEW
OF THE COMPARED ATLANTIC
ALPHABETS & GLYPHS
OF AFRICA AND AMERICA.
By Prof. C. S. RAFINESQUE. Philadelphia. 1832.

LYBIAN.
1. Primitive and Acrostic.
2. Old Demotic or Tuaric.
Meanings and Names of Letters in No. 1.

AMERICAN.
3. Letters of Otolum.
4. Glyphs of Otolum.
Names of Letters in No. 2.

ATLANTIC JOURNAL,

AND

FRIEND OF KNOWLEDGE.

IN EIGHT NUMBERS.

CONTAINING about 160 original articles and tracts on Natural and Historical Sciences, the Description of about 150 New Plants, and 100 New Animals or Fossils. Many Vocabularies of Languages, Historical and Geological Facts, &c. &c. &c.

BY C. S. RAFINESQUE, A. M...PH. D.

Professor of Historical and Natural Sciences, Member of several learned societies in Europe and America, &c.

Knowledge is the mental food of man.

FIGURES.

Melissa or Balm.	page 14	Tubular shell,	page 127
Mammoth Cave,	27	7 New Fossil Shells,	142
Franklinia,	79	American and Lybian Glyphs	
Fossil Teeth,	100	or Primitive Alphabets,	38

PHILADELPHIA:
1832—1833.
(TWO DOLLARS.)

Frontespiece from Rafinesque's *Atlantic Journal*, 1832–1833. The chart created by Rafinesque compares and links the languages of America and Africa. Library Company of Philadelphia.

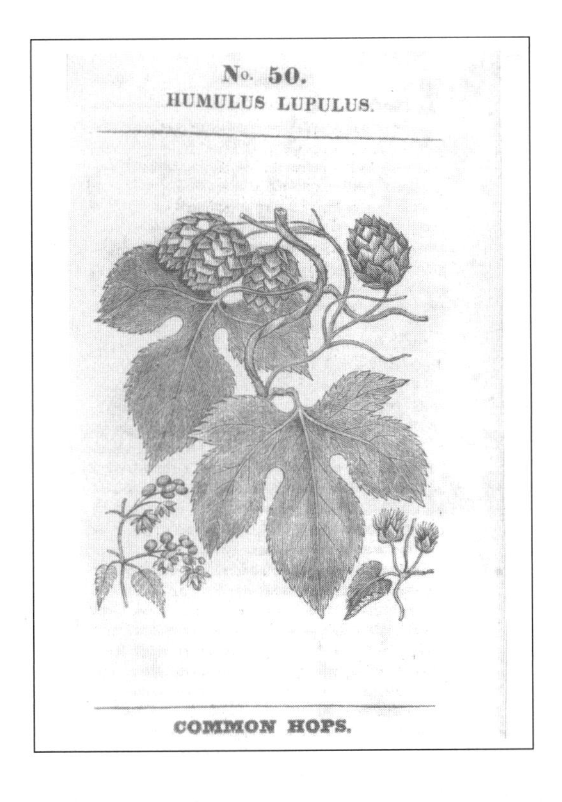

Woodcut based on a drawing by Rafinesque for his *Medical Flora*, vol. 1, 1828. Van Pelt Library, University of Pennsylvania.

Wallam Olum

First & Second Parts of the
Painted and engraved traditions
of the Linnilinapi —

II Part

Historical chronicles or Annals
in two chronicles

I. from arrival in America to settlement
in Ohio &c 4 chapters, each of 16
Verses, each of 4 words. 64 Signs

2d From Ohio to atlantic States &
back to Missouri, a mere Succession of
names in 3 chapters of 20 Verses 60 Signs

translated word for word by means of Zeisberger
and Linapi Dictionary. with explanations &c

By C. S. Rafinesque 1833.

Title page of *Walam Olum*. The original Rafinesque manuscript is in the Annenberg Rare Book and Manuscript Library, Van Pelt-Dietrich Library Center, University of Pennsylvania.

47. Opekasit — Easterly looking 46
Sakimanep King was.
Sakhelendam — being sad
pallitonepit — at the warfare
48. Wapagishik — East Sun or Sunrise
yuknohokluen Let us go saying
Makelohok they are many
wapaneken — East going together

49. Tschepicken — Separated
Namaesipi . Fish River
Nolandowak lazy they
gunehunga — they tarry
50. Yagawanend — Hut maker 47
Sakimanep King was
Talligewi — Talegan or there found
Wapawullaton — East possessing
51. Chitanitis — Strong friend 48
Sakimanep King was
Wapawaki — East rich land
gotatamen he desires.
52. Wapallendi — East some
pumisinep — went or passed
Talegawil Taliga head or Emperor
Allendhilla — Some kill
53. Mayoksuwi — of one mind
Wemilowi all say

Verses of Book IV of the *Walam Olum* with associated pictograms and translations, from the Indiana Historical Society publication, 1954.
Verse 47. When Opossum Face was chief, he worried about the destruction of things belonging to others.
Verse 48. Now when daylight came, he spoke three times: 'Let those going east be many.'
Verse 49. They separated at——? River; and the ones who were lazy returned to Snow Mountain.
Verse 50. When Lean-to Man was chief, the Talligewi were in possession of the least.
Verse 51. When Strong Friend was chief, he wanted to go to the eastern country.
Verse 52. When some infiltrated into the east where the Talligewi were, some were killed.
Verse 53. In right-minded indignation, all said 'Let us despoil! Let us destroy!'

Rafinesque's tomb in Old Morrison, Transylvania University.
Photograph from the Transylvania University Library.

ing Rafinesque to comment: "this kind of merit and lofty knowledge is so little understood and valued here, that some periodicals have refused even to notice this literary fact!" To add to the misfortune, of the one thousand francs offered as part of the prize, he received only one hundred, and a promise of publication of the essay was never honored. Still, he was happy to receive tangible evidence of appreciation for his effort—bursting with pride, he announced his triumph to the world.[9] As with virtually all of his historical and ethnographic speculations, his claims have not survived the test of time, and indeed, they would seem to be preposterous.

In contrast with Rafinesque's experience with the French judges, he was embittered by his abuse at the hands of American critics. In 1825 he felt he had "won" the two American contests he had entered but never received a prize. One essay was for a prize of $1,000 offered by the federal government for the most effective means of clearing the Ohio River of snags and trees; the prize went to a contractor. The other essay was for a prize of $100 for the best paper on Indian tribes and was offered by the Academy of Science of Boston. Though his essay was deemed the finest, the prize was not awarded because the work was considered too long.

Perhaps as an outsider by birth, and by nature, Rafinesque felt a kinship with blacks, gypsies, and Jews. He assumed the responsibility of providing the public with authoritative facts about them, setting right any erroneous statement he happened upon. Ever the letter writer, ready to comment on any matter, he responded to an article in Silliman's Journal that claimed there were no Gypsies in America. He corrected this error, declaring that early on they had been sent to Brazil and the Argentine by Spain and Portugal, but that most (five million) remained in Asia, Africa, and Europe.[10] He published statistics about Jews, providing a country-by-country breakdown of their distribution and the languages spoken—in 1829 there were 2.7 million Jews in the world, 67 percent of whom were in Europe, and 1.1 percent were in America.[11]

Rafinesque had created many new species and genera of plants on the basis of geographic location and the slightest difference in their form. Yet, despite all the differences found in humans, which he assiduously described and analyzed, he was a unifier or a "lumper," insisting that humans were all of one species, deeply intermixed culturally and biologically—a conclusion dictated by his humanistic philosophy.[12]

Since the origins of humankind lay somewhere in Asia, according to the Bible and other writings, by what routes did migrations take place to America and when did they occur? Speculation on these questions began within a few years of Columbus's discovery of America and has continued to this day, attracting as it did mystics and crackpots, as well as sober scholars. European newcomers were confronted by two apparently disparate groups of inhabitants who had arrived ages ago—the Native American, the "red Indian," a primitive people who came from Asia across the Bering strait, and the supposedly more ancient, and far more sophisticated inhabitants of Central and South America and Mexico who had left behind great stone monuments bearing hieroglyphics. Very few who had thought about aboriginal Americans believed that the New World was their site of origin, because humankind had been created elsewhere, and since the very beginning all Americans have been deemed immigrants.

The history of aboriginals and the enumeration of new American plants and animals were of far greater relevance to the lives of early nineteenth-century Americans than were theoretical physics and chemistry and would be better left to the Europeans. Americans could seize upon the advantage of having at hand plants, animals, Native Americans, and relics of ancient civilizations. But little reliable knowledge about the past existed, and the methodology for acquiring this knowledge was not yet developed. Only after years of effort was a conceptual framework for reconstructing the events of earlier times worked out. Hard "facts" were scarce, most coming from scripture, legends, natural science, and historical and geographical accounts. Some historians and philologists, including the inventive Rafinesque, were unfettered in their historical narratives, and so the literature abounded in conflicting stories, some reasonable, others absurd, and all almost always wrong.

In the course of world history, civilizations had come and gone, leaving magnificent reminders of their brilliance in Greece, Italy, Egypt, Crete, Cambodia, Mexico, and Central and South America, but in North America the mounds found scattered about were primitive, often clumsy, and overgrown with vegetation. They were a mystery, although it was generally felt that they were associated with defense, death, and burial. One can only imagine the puzzlement of Americans confronted by the mounds, fertile soil for the growth of theories and half-baked notions about earlier inhabitants. The systematic study of great, lonely mounds scattered over the Mississippi Valley and the Southeast was prompted by Thomas Jefferson's fascination with them in Virginia. Jefferson's study of these relics, methodically excavating the mounds layer by layer, added to the archaeological record as he probed deeper and deeper into the past.[13]

Native Americans of the region could not explain the presence of mounds or determine who had built them, and this information was not in their legends. Some "excavations" by treasure hunters and graverobbers provided information, but most studies by travelers and natural historians such as William Bartram and Rafinesque, simply described and mapped the monuments and tried to explain how they fit into aboriginal history, such as it was known and understood.[14] To Americans, one of the compelling questions that systematic investigation might answer was the origin of the Native American and of humankind itself. Secular scholars, who were convinced that comparative linguistics was the instrument by which languages could be traced back to a common root in the distant past, wanted to demystify the problem through observation and reasoned analysis. They felt that comparative studies in ethnology, physical anthropology, and archaeology would provide insight into the histories and cultures of the various American nations. The strategy would not only provide the best opportunity for determining the ancestry of the Native American, but some believed it might also enable scholars to keep going further back in time until the progenitor human could be identified, a remote goal that presupposed all human languages were related and that somehow information about extinct languages could be recovered. The subject was of intense interest to both religious and secular scholars, each confident that the burning question of whether the creation of humankind occurred once or on several occasions would be answered in a scientific manner. The religious philologist was anxious to validate the biblical version of Creation while the Enlightenment humanist wanted to discredit it.[15]

Jefferson,[16] along with others such as Benjamin Smith Barton[17] and Rafinesque, was zealous in compiling lists of words from native American languages, some already published, others provided by a network of people who dealt with Native Americans, including Indian agents, travelers, soldiers, and priests. Jefferson had instructed Meriwether Lewis to record the vocabularies of the various tribes he came in contact with (Mandan, Sioux, Shoshone, Nez Percé, and others) during his expedition with Clark to the Pacific Ocean (1804–1806); this he did assiduously, collecting nine Native American vocabularies. Aware of the fate that faced the Native American, he was eager to record what he could of their languages, history, and culture before they were lost forever.

Rafinesque was at a crippling disadvantage compared to Jefferson and Barton, for he could never command the extensive word-gathering resources that they could, but what he lacked in information he made up for in energy and in imaginative reconstructions of history and language lineages. In fact,

Rafinesque was better equipped to study language, comparative philology, and ethnology than any of his contemporaries because of his familiarity with history and several languages, both modern (English, German, French, Spanish, and Italian) and ancient (Hebrew, Greek, and Latin).

Words were compared and classified in order to establish relationships between tribes and to determine affinities between them, which is akin to the procedure used in botany; Rafinesque's ambition was to become the Linnaeus of both botany and philology.[18] From the fraction of words common to different languages, scholars could establish the closeness of two languages to each other, which permitted the identification of a language as a parent to others, or as a sibling, so that it might be possible to trace various dialects back to their mother tongue, thus forming a linguistic tree. Rafinesque insisted that words, mainly of objects and adjectives, were more important in establishing an affinity between languages and dialects than was grammar—a modern view. A simple method was devised by Rafinesque for expressing affinity *numerically* by his *Synoremic formula,* and although the method was crude, it identified Rafinesque as one of the first lexicostatisticians with a semi-quantitative method for the determination of the affinity of languages and the ancient relationships of different ethnic groups.[19]

Confidently he wrote that he wanted to bring extinct languages to life. "I take scattered words of extinct nations and Languages, and out of a few or any number, I restore them to our historical knowledge. . . . Give me but a single *genuine word* of an ancient or extinct Language, and I can find out its analogies with all others. Give me 2 or 3 or a few, and I can trace its alliances. Give me several, and all its origins, parentage, filiation, claims, affinities, peculiarities, &c can be traced." With Champollion's triumphant decoding of hieroglyphics ringing in his ears, Rafinesque was overly optimistic but he most certainly was on the right track, even at this early date.[20]

In histories of American linguistics, Rafinesque's name is nowhere to be found. Yet according to the modern linguist Belyi, Rafinesque can be credited with providing remarkable insights into the fundamental nature of human language. And Belyi considers Rafinesque "one of the founders of the scientific study of American Indian languages." If he made mistakes ("far-fetched and wrong" in some of his writing), they were no worse than those of his contemporaries, and he should not be judged by modern standards—those reexamining his botanical classifications made a similar plea. Rafinesque also studied the *written* language of aboriginal people, and as a born classifier he described twelve kinds of writing, pointing out that abstract ideas could be expressed by aboriginal language and their symbols—a notion that was contrary to accepted belief.[21]

The first four decades of the nineteenth century—Rafinesque's time—witnessed rapid growth in the study of Native American languages, progressing from the basic collection and comparison of vocabularies of different languages to the more sophisticated study of the sound of words and the establishment of a system of phonetic symbols that could be understood by all scholars. Rafinesque can be credited with the call for a universal system of describing sounds, a phonology, and an orthography that would facilitate the comparative study of languages, both of the Old and the New Worlds. A massive five-volume work, *Synglosson*, in which he proposed to record the vocabularies from America, Asia, Africa, Europe, and Polynesia, never came into existence.

With the involvement of German scholars, such as Alexander Humboldt and Johann S. Vater, a note of caution was introduced in the enthusiastic establishment of affinities between Native American tribes and nations proposed by Barton and later by Rafinesque. There simply wasn't enough information, and what information did exist was too scattered. Major contributions to comparative linguistics and to the understanding of tribal relationships were made by several American scholars—Peter S. Duponceau, John Pickering, John Heckewelder, and especially Albert Gallatin, whom the Europeans regarded as having made a sound and original American contribution to world culture.

Rafinesque's first work in Indian linguistics dated from his time in Kentucky, and after a decade or so of study he concluded that the number of Indian languages was manageable and not beyond human understanding, "reducing the 1,800 American Dialects to about 25 Generic languages . . . 14 from North and 11 from South America." He listed each of these, complete with the number of dialects found within a particular generic language. For instance, under the generic language of *Nachez,* he claimed to have identified nearly seventy-five dialects and tribes. Rafinesque wove components of the twenty-five generic languages into a tissue of impossible detail that he claimed would be "the key to American Ethnology, Philology and History," all due to his research whose proofs, he stated, "would fill volumes," none of which ever appeared.[22]

Most of his philological efforts were directed toward the languages of the New World, but he also delved into others,[23] for he began with the premise that *all* languages were related. The bias can be seen in a study comparing

English with Coptic. "From 252 Coptic words, collected at random for comparison, I find 83 more or less alike with the English." Perusal of the word lists suggests that Rafinesque's criteria were highly questionable or inadmissible.[24]

A new phase in Rafinesque's research on language began with the publication of an open letter in the *Saturday Evening Post* to Peter Stephen Duponceau, a Philadelphia lawyer of French birth and a talented linguist whose work on the structure of Native American language had earned him international respect. Apparently he also impressed Rafinesque, although the two philologists seemed to have had little direct contact with each other despite living in the same city. Rather than post a letter to a correspondent, Rafinesque would frequently publish it as an open letter if the addressee were well known (Duponceau, Champollion, McCulloh) and the subject was of general interest. Rafinesque wanted the widest possible audience for his thoughts and would not let any of his writings go to waste.

Inspired by Champollion's epochal work decoding the hieroglyphs of Egypt, Rafinesque's landmark letter is the first recorded attempt to interpret ancient Mayan hieroglyphics. The glyphs had come from the ancient city of Otolum in the Yucatan and were distinct from Aztec, Mexican, and all other written letters that he considered true symbols. Astutely, he asserted that ancient Mayan writing was related to the modern Mayan language and that numbers were designated by bars and dots—"numbers are perspicuously delineated by long ellipsoids marking ten with little balls for unities, standing apart." Symbols were grouped in syllables, words, or short sentences. The value of this important communication was marred by speculation that was presented as certainty about the relationship of Mayan glyphs to the "Lybian" and north African scripts, and the linkage of the Otolum Empire, founded by a branch of the Atlanteans, with the Etruscans, Egyptians, and Persians.[25]

Rafinesque's philological papers, especially the open letter to Duponceau, must have attracted some attention, for it initiated a correspondence with James H. McCulloh, a customs collector from Baltimore and a graduate of the University of Pennsylvania School of Medicine. An amateur historian, consumed with an interest in the early inhabitants of America and eager to learn, his contact with Rafinesque resulted in *Four Open Letters on American History* to McCulloh published in the *Saturday Evening Post*.[26] The letters contained an overview of human racial types, always emphasizing the unity of humankind despite the variety of features found in the various parts of the world. Rafinesque claimed that nations and tribes of every color were in America before the arrival of Columbus and that humans of every hue could also be found in Asia, Africa, and Polynesia. Rafinesque reiterated his conviction that the origin of American nations and tribes was Atlantean, (before Atlantis sank under the waves). Believing that "all languages may be traced up to one or few original stocks," he made many astonishing claims, including the overlapping of the Carib language with Hebrew. Rafinesque claimed that his understanding was different and more profound than that found in standard histories because of his use of sources not consulted by North American historians, but in truth his fabulous conclusions were the fruit of his untrammeled imagination.

McCulloh, the student, and Rafinesque, the master, became friends, freely communicating with each other. The student was wise and stable, able to select sound information and theory from his teacher's emanations. McCulloh acknowledged his debt to Rafinesque in the preface to his *Researches, Philosophical and Antiquarian, Concerning the Aboriginal History of America* (1829), a valuable work that contained a sound discussion of Mayan hieroglyphics free of unsubstantiated speculation.[27]

The publication of Rafinesque's historical and philological papers in popular magazines such as *The Saturday Evening Post* and *The Cincinnati Literary Gazette* seems odd. These journals were business ventures striving for a broad readership to increase profitability, and yet publishers were willing to print detailed accounts of a technical nature that belonged in specialized professional journals, scarce as they were at the time. Scientific journals were closed to Rafinesque, so in effect he was going over the heads of a small, highly critical, "professional" in-group of his peers to seek a broad audience, learned or not. To enhance the interest of his writings, they were frequently enriched with colorful facts and intriguing theories woven into a narrative that would impress and appeal to a less critical readership. In the first third of the nineteenth century, articles such as these were an important pop educational force, catering to people who wanted to "improve" themselves, just as today the Science Section of *The New York Times* is read by a public thirsting for comprehensible scientific knowledge. There lurks the suspicion however, that editors exploited some of Rafinesque's material as humorous caricatures of the so-called learned treatise by ivory tower professors in a practical age that was not without its anti-intellectual sentiments.

Rafinesque must have been aware of the fact that he was rarely quoted. Serious scholars who knew something about the field and how difficult it was to establish the most elementary relationship in linguistics ignored what he had to say, although not infrequently it was something of value. But how could the valuable and the nonsensical be sorted out? Peter Duponceau made it no secret that he ignored Rafinesque, and as an influential Philadelphian and president of the American Philosophical Society, made sure that Rafinesque remained a scientific outcast, excluded from the Society—a possible case of a respectable French émigré loudly dissociating himself from a disreputable confrère. However, Rafinesque made full use of the society's library and archives. If Rafinesque had any impact in the field, it was silent and unrecorded. The general public might be impressed by the serious, narrative presentation with troves of names, and their relationships, but experts read and repudiated them.[28]

With their faith in progress and improvement, citizens of the Republic found it impossible to believe that a once proud and advanced civilization that could create great mounds and the impressive monuments of Mexico and Central and South America could "degenerate" to the primitive state of the Native Americans in their midst. Those who left ancient buildings and monuments would have to be of separate origin and not the progenitors of present-day "savages." Few believed in the single immigration theory. Rafinesque and others felt that the Mound Builders were driven south by a second wave of immigrants, cultural inferiors from Asia. According to this scheme, the earlier groups, the superior Atalans from Europe and North Africa, were the Mound Builders of the Midwest, perhaps indulging in architectural exercises preliminary to the building of the great, monumental structures of Central and South America, where they had been driven by the later group of Asians, ancestors of the Native Americans.[29] Others believed that the mound builders had come from across the Pacific Ocean.

Numerous commentators eagerly sought evidence for their own version of migration, one more fantastic than the other. One historian claimed that Brazil was populated by Carthaginians and Israelites in the time of Solomon, about 1000 B.C.; another believed that the very earliest Americans were Scythians, the descendents of Shem who also gave rise to the Jews. There were remarkably learned theories that involved Malays, Polynesians, and Australasians from the west, Atlanteans from the east, with additions throughout ancient history of Egyptians, Syrians, Europeans, Greeks, Romans, and others. Josiah Priest, an industrious but uncritical amateur scholar who borrowed freely from others, solemnly stated that "of the twenty five original nations of America, three were Atalantans," and he came to believe that Noah's three sons were each of a different color—black, white, and red.[30]

Savants, including Jefferson, were struck by the physical resemblance between Native Americans and Chinese and considered them of Asian origin, descending from the Northwest. One of many commentators, Samuel L. Mitchill, who believed that the Garden of Eden was in North America, created an elaborate story of early North American peoples—that the mounds of the near Midwest were built by Malays. Jefferson at one point came to the conclusion that the people of the New World were of greater antiquity than those of the Old. Questions were asked, many theories were proposed, and data was gathered as the methodology in archaeology and linguistics improved.

Reviewing the extant literature, Rafinesque concluded that the single cradle of humankind was in the central high country of Asia—the "Imalaya mountains," from which humans spread to all parts of the globe taking with them animals, fruits, and plants. To support his thesis, he asserted (erroneously) that "All our fruit trees, all our cereal plants, and nearly all our culinary plants are also found growing wild in those mountains." The major human group that radiated from the Himalayan area was the *Atalantes*, which conquered and occupied Europe and Africa, spent some time on the "Atlantic islands" (Atlantis), and eventually went to America. Rafinesque *proved* that it was from this stock that the American nations arose. Their language, according to Rafinesque, gave rise to the names *Atlas* mountains, *Italy,* (Aitala), and *Atlantic,* as well as many other European and north African place names. Typically, his reviews teemed with lists of names—tribes, languages, and places—that taxed the reader mightily, but surprisingly, he convinced the editor of the Saturday Evening Post to allow publication of his wilder recitations.[31]

The actual historical record was sparse, and whatever there was had survived over thousands of years in colorful but dim legends embellished by clever storytellers. In the absence of hard data, gratuitous assertions, theories, and opinions abounded—and became *fact*—almost all of which have since been discredited. However, the comparative study of language and customs became a constructive and fruitful enterprise to establish relationships between human groups. With some justification, Rafinesque has been credited with having a dynamic, evolutionary conception of language, but how he established his classification of myriad languages was beyond what was known and was beyond comprehension.

In 1819, four years after his return to America, Rafinesque's first paper on Indian archaeology appeared. In the paper he described an "ancient monument or fortification," an earthwork at Chillicothe near Lexington, which was later found to be at least nine hundred years old. Rafinesque believed it to be the remains of a walled town of the Alleghaween nation, which he investigated in some detail—their "acquirements and their civil history." Soon he described several other Alleghawian monuments in northern Kentucky and made the point that they were similar to the remains of the ancient Floridian, Antillian, Mexican, Peruvian, and Chilean nations. Whether earthworks were thrown up for defense or for religious and burial purposes was open to debate. A global thinker, Rafinesque saw similarities between Alleghawian remains not dedicated to defense and "many Celtic, Druidic, Scythic, Tatarian, Indian and Polynesian religious monuments."[32] The work was not unlike his botanizing, for he explored outdoors, pacing off distances, covering grassy fields back and forth, looking for ancient human artifacts, and probably plants at the same time. Much to his pleasure, the work was solitary—Rafinesque alone, at peace with the world, communing with heaven and earth—and in this best of worlds he had ready access to publication in the *Western Review and Miscellaneous Magazine,* a monthly journal sustained by John Clifford.

Clifford had kindled Rafinesque's interest in ancient mounds and their builders and was his mentor on the subject. Rafinesque's thinking about the Asiatic origin of Native Americans stemmed from Clifford's notions of their Hindu origins. The two men were remarkably compatible, with the younger man doing the legwork and collecting the data. Unfortunately, their productive partnership suddenly ended in 1820 with Clifford's death.

Rafinesque's numerous descriptions and sketches of Indian mounds were sent to his colleague Caleb Atwater and published as open letters in the *Western Review and Miscellaneous Magazine.*[33] Atwater, postmaster of Circleville, Ohio, and an attorney, was dedicated to the study of near midwest, prehistoric Americans, and had a more comprehensive understanding of the subject than most because he corresponded with anyone who had anything to say about the subject. Operating under the auspices of the American Antiquarian Society of Worcester Massachusetts, he had become a clearing house for information, and in 1820 he published *Description of the Antiquities Discovered in the State of Ohio and Other Western States,* the first major archaeological work in America west of the Alleghenies, but it was a work that frequently went beyond what was actually known.[34]

By the time of publication, Rafinesque was no longer on cordial terms with Atwater because he felt Atwater did not give him and others adequate credit for their findings. (Indeed, Atwater never mentioned Rafinesque). The book was, in general, well-received, but Rafinesque's critique in his *Western Review and Miscellaneous Magazine* was, as expected, both friendly and distinctly unfriendly. Tentatively, there were some areas of agreement between Atwater and Rafinesque. The latter commented sympathetically on the Mosaic account of Creation, favored by Atwater, but reserved judgment on Atwater's view that the mounds were not built by the ancestors of present day Native Americans but by the Jews, because the nature and hill locations of American mounds bore resemblance to those of the Middle East. He felt there was much to criticize in Atwater's acceptance of the evidence for an early European presence in America, and he derided Atwater's belief that the discovery of Roman coins found in a cave in Nashville, Tennessee, was valid. John Haywood, who first described the Roman coins, also believed that the Ohio and Mississippi valleys had been the home of an invading race of giants.[35]

Rafinesque praised the book as a "valuable compendium . . . an excellent supplement to the works of many others. . . . [It] contains a mass of important information, deserving the attention of the antiquary and the man of science . . . without displaying any great originality of thought or laying claims to any peculiar merit for excellence of arrangement or perspicuity of method." Rafinesque was "amused and instructed" by the memoir, the style being "animated," but it was "diffuse," and "not exempt from grammatical errors." Rafinesque was offended by Atwater's pose that he wrote his book after great personal research and labor, contemptuous of "crude and indigested statements of travellers." Atwater had, in fact, "adopted" the data of many—"Mr. Clifford, Dr. Mitchill, Rev. Mr. Heckewelder, Professor Rafinesque and others . . . [shamelessly omitting] to mention to whom he is indebted for [his] particular views." Rafinesque faulted Atwater for the incompleteness and inaccuracies of his work. Though he admired Atwater's "zeal and industry," he felt that "if he could have studied more and attempted less . . . the results would have been more beneficial to history and science."[36] Despite its shortcomings, and Rafinesque's carping criticism, Atwater's book was a seminal, compendium of Indian mounds and their contents, indispensable for future studies.

His life's work trashed, Atwater was infuriated by "that Italian," whom he called a "pretended professor . . . the worst of impostors, in literature and science, now living in the world."[37] Rafinesque had added to his list of enemies, but this one was different. Rather than fume in silence, he began a campaign of vilification to destroy Rafinesque's reputation.

In view of Rafinesque's brilliance, his encyclopedic knowledge, and his essential generosity and good will, the inordinate degree to which he was reviled has always been perplexing. He did have admirers, but they were less influential than his detractors—Gray, Haldeman, Lea, Binney, Harlan, Short, Drake, Duponceau, and others—who were not silent, and the scientific world took its cues from them. Boewe has advanced the plausible theory that Atwater's vigorous enmity fueled the movement to reject Rafinesque. Taking full advantage of his position as postmaster (free postage), he sent out defamatory letters that may have been responsible for Rafinesque's consistent lack of success in obtaining jobs and publishers. Letters from Atwater to Benjamin Silliman seem to have been partly responsible for the abrupt cessation of Rafinesque's papers in Silliman's respected *American Journal of Science*. A letter to Isaiah Thomas of the influential American Antiquarian Society had an equally harmful result, for many of Rafinesque's manuscripts containing important archaeological data were received but never published in their journal; the manuscripts were discovered in the society's archives many years later. Upon receipt of a damning letter, Horace Holley of Lexington wisely requested evidence for Atwater's violent denunciations. There seems to be little question that this far-flung fulmination assured the deterioration of Rafinesque's status, and though it was not the sole impetus, it helped nurture hostility against Rafinesque for half a century.[38]

Rafinesque's *Ancient Annals of Kentucky* (1824), dedicated to Humboldt, summarizing the first phase of his studies, was published with "philological and ethnological tables abridged from an elaborate survey of about 500 language and dialects, reduced to 50 mother languages with principal roots for four important words," namely, *heaven, land, water,* and *man.* Although Rafinesque could be highly critical of claims that provided scientific support for some aspects of the biblical story, he often linked historical events with biblical accounts. He described the stages in the early geological history of the earth beginning with the great flood and its aftermath (complete with biblical quotes), correlating various stages of the Mosaic account with his own six periods of formation of the soil of Kentucky, the last stage witnessing awful volcanic eruptions of the sea, convulsing the Atlantic ocean, with monstrous mammoths, elephants, and mastodons roaming over the land. Rafinesque firmly believed that "there was regular intercourse between all the primitive nations from the Ganges to the Mississippi," and he proposed that before the Christian era, settled nations were displaced or annihilated by invading nations after titanic battles, that the Phoenicians had traded with America, and that Etruscans wanted to settle in America but were prevented by the Carthaginians.

At the end of *Ancient Annals,* he apologized: "All details which might have explained, and notes that would have proved my statements are unavoidably omitted" due to a lack of space, and he implied that they would soon be published. They never were, but vast amounts of information and theory were scattered in various publications over many years—often published twice.[39]

Having condensed the detailed history of the world from its origin to the late eighteenth century to less than fifty pages, he most certainly lacked the space.

To Rafinesque's credit, despite all the shortcomings, the *Ancient Annals of Kentucky* was the very first accounting of most of the antique, Native American monuments in Kentucky and its neighboring states, a work of great use to future workers. Still, D.G. Brinton, perhaps the leading American philologist and archaeologist of the late nineteenth century and an admirer of Rafinesque, called *Ancient Annals of Kentucky* an "absurd production, a reconstruction of alleged history on the flimsiest foundations." But then he concluded, "But, alas! not a whit more absurd than the laborious card houses of many a subsequent antiquary of renown." *Absurd* was Rafinesque's favorite word.[40]

In response to a critical review of his work, Rafinesque wrote a lead article in the *Cincinnati Literary Gazette.* "[A] review has appeared which should have only excited sentiments of contempt and pity, for the ironical praise, hints and ignorance of the writer,—if many of his statements, or inuendos, had not the obvious tendency to distort, depreciate and mutilate some of the historical facts and statements which I have evolved from assiduous researches and labours."[41] His response was lengthy and elaborate, but there were few who took it seriously. On the page following his defense of *Ancient History of Kentucky* there is a mock serious letter to the editor berating Rafinesque for announcing to the world that he had made three discoveries of staggering importance while denying it any details. The writer of the letter pleaded with Rafinesque not to withhold his secrets for the good of humanity.

Rafinesque must have had a remarkable hold on the editor of the *Cincinnati Literary Gazette,* for almost certainly a lengthy, unsigned critique exulting over the work was written by Rafinesque himself.[42] The style, the details emphasized, the sharp criticism of Atwater, and the affable disagreement with the influential Dr. Mitchill of New York would suggest that this was so, despite a few odd, incorrect statements that Rafinesque would have known to be untrue, perhaps a feeble attempt to disguise his identity as author of the review. For instance, it was stated that he was a native of Greece, a misunderstanding that Rafinesque, at the time a public lecturer on the war of Greek independence from Turkey, would have wanted to cultivate, for a wave of sympathy for the beleaguered Greeks was sweeping America. Rafinesque was extravagantly praised in terms precisely similar to those he attributed to himself: "a production that carries with it the impress of *genius* . . . that distinguished philosopher, Professor Rafinesque whose very name is synonimous [sic] with literature and science. There is perhaps no man living, the aggregate of whose knowledge, equals that of the worthy Professor. There is no ramification of knowledge with which he is not familiar. . . . He is revered as the great storehouse of knowledge, and his name is already written on the scroll of fame." No one capable of reviewing such a book in detail would have praised him so; the judgments of the reviewer (Rafinesque), revealed how he wanted to be judged by others, how he craved recognition by both the public and his intellectual peers. In his opinion, his accomplishments in linguistics and history were the equal of the renowned Champollion, and yet he felt they remained unacknowledged and unrewarded. On the other hand, the article could have been an elaborate send-up. If he were indeed the author of this panegyric, it was an infantile, shameless exercise, whose origin lay in delusion and psychopathology.

Rafinesque's *The American Nations . . .*, published twelve years later, after numerous promises and announcements that its appearance was imminent, was another vastly ambitious work that taxed one's credulity. It was to be published in six volumes but only the first two were printed, each about three hundred pages and costing one dollar. The compass of the book was revealed in the subtitle: "the whole history of the earth and mankind in the Western Hemisphere; the philosophy of American history; the annals, traditions, civilization, languages &c of all the American nations, tribes, empires, and states." He began his story with the Creation, 6,690 years before Columbus, and concluded with accounts of the Lenape and other nations, but despite its claims, the work contained little that was new. One critic dismissed it with: "Its pages are filled with extravagant theories and baseless analogies."[43]

The preface and introduction were rather combative: "All the histories of America [hitherto published] are mere fragments or dreams." To Rafinesque's credit, he brought attention to the incompleteness of early histories of America by English-speaking historians, pointing out the important work of Spanish historians and missionaries that hadn't been translated into English, and whom Rafinesque regarded as "the fathers of our history." Meanwhile, because of the "indolence of our historians," valuable sources of information were neglected and old errors and myths were propagated because original sources were not examined. He assumed the responsibility of writing such a work because previous works in English were "paltry" and without style,[44] and his demand that historians broaden the range of their sources and compare the languages, religions, and customs of nations in all parts of the world could not be faulted.[45]

In 1856, Haven, reviewing Rafinesque's contributions to American archaeology, noted that the details in his accounts were so complete "as to leave little to be desired in the way of precise information." His writings were a "succession of peoples and empires, with a lavish profusion of names and pedigrees, and an air of intimate acquaintance with their civil and religious customs, and the motives and results of military operations, which seem to imply the possession of an insight the reverse of prophetic, but equally supernatural." The only thing the Rafinesque lacked was "the faculty of judicious discrimination."[46] The critique was a reasoned assessment and disposition of Rafinesque's version of history—and terribly damning. Published not long after his death, the view expressed by Haven was undoubtedly that held by most serious students.

Throughout the span of historical study of the New World there have been historians dedicated to the belief that the Hebrews of the Old Testament were early colonizers of America—perhaps the original immigrants. In the absence of firm knowledge, the imagination could run wild, so that the reconstructions of the story were little more than literary exercises. The English settlers of Massachusetts and the Mormons were convinced that the North American Indians were one of the lost tribes of Israel, a belief held by William Penn, Roger Williams, and Increase and Cotton Mather.

Josiah Priest summarized current information on the subject in his *American Antiquities and Discoveries in the West,* an undisciplined work that was widely read, in which he forcefully argued that the Indians were the lineal descendants of all ten tribes, carried in captivity to America 2,500 years ago. Kinships were discerned between the Indian and Hebrew languages, both groups were supposedly monotheistic and practiced circumcision. Both also washed and anointed their dead before burial. The Old Testament was scoured for supporting evidence—and the evidence was found! There was a general sympathy for the view that the Bible was in essence true, and the task of the righteous historian was to reconcile the biblical record with the secular and scientific findings of the day.[47]

On more than one occasion Rafinesque expressed a respectful interest in biblical interpretations and, along with others, was convinced of a remarkably free migration of peoples between continents.[48] However, contrary to notions periodically revived by fervid religious factions, Rafinesque insisted that there were no Jews in the New World before Columbus and that Native Americans were not descended from the ancient Hebrews. He wrote a harsh review of Joseph Smith's *Views of the Hebrews,* (1823), describing Smith's elaborately argued notion of Indian descent from the Hebrews as a "singular but absurd notion," a view "based upon some religious prejudices and slight acquaintance with philology and antiquity. . . . It is to me astonishing how in this enlightened age, any such unfounded belief can be sustained." The eight "proofs" supporting Smith's theory of Native American origins were examined and effectively destroyed by Rafinesque, point by point, and then he presented ten learned and devastating arguments to prove the theory false and unacceptable, a position on the subject that was unequivocally stated in the title of his article, "The American Nations and Tribes are not Jews."[49] He deplored the fact that the Mormon faith had incorporated a naive theory of Native American origins that had become encoded in the *Articles of Faith for the Book of Mormon.* He also lamented the folly of Edward King, Viscount Kingsborough, who impoverished himself and died in a debtor's prison after spending at least £80,000 producing a seven-volume work supporting the idea of the Israelite origin of the Mexican Indians.

Two years before his death, Rafinesque summarized his thinking about the Native American in his *The Ancient Monuments of North and South America*, a remarkable document that embodied the Rafinesquian paradox. On the one hand, it was a seemingly balanced and erudite assessment of the state of knowledge of the field, rightly criticizing the "egregious mistakes" made by some historians who indulged in "dreams of systems based on a few facts" that could easily be "overruled by hundreds of facts." He discussed some of the most "absurd" examples and provided arguments that eliminated them from the realm of possibility, but no sooner had Rafinesque finished with the faulty notions of others, then he proceeded to espouse his own questionable and unsupportable theories. The disparity between the reasoned and discerning critiques of others' works and the wildness of his own creations is truly remarkable.

His detailed criticism of others was so closely applicable to his own shortcomings that one wonders whether in some twisted way he was writing about his own work. His reference to the misleading "dreams" of historians might have reflected his own dreamlike, delusional state in which fantastic stories, not unlike Baron Munchausen's, became realities. It would seem that his critical faculty about himself had ceased to function, for he was almost impervious to criticism, suggestion, and advice.

In the end, in a field of endeavor where facts are hard to come by, and (often) one opinion is as good as another, we are witness to the battle that raged in Rafinesque's head, between his intelligence, broad knowledge, and great power of reasoning on the one hand, and his tortured, distorting imagination on the other. In Rafinesque's frequent times of stress, the latter seemed to gain the upper hand, leading to public displays that seemed preposterous to reasonable people, and this was a pattern that repeated itself, whatever the field of interest. Where a line could be drawn between the rational and the irrational was a matter of debate between people who admired or detested him.

Chapter 12

WALAM OLUM

The saga of the Lenni Lenape or Delaware Indians, their epic wandering from Asia to the shores of the Delaware River, was recorded in the *Walam Olum* ("Red Score" or "Painted Sticks"), an ancient document of immense importance, if it proved to be authentic. "Discovered" and translated by Rafinesque, its true origin was an enigma, and even the authenticity of the document itself was challenged by skeptics for more than one and a half centuries. The chronicle encompassed nothing less than a Creation myth, a flood legend, the entry of the tribe from Asia to Alaska, their migration to the south and east to the Delaware/New Jersey/eastern Pennsylvania area, and a chronology of ninety-six successive chiefs, all unfolding over 3,200 years—an unbelievable story.

According to Rafinesque, while roving about Kentucky botanizing and examining Indian mounds, he befriended a "Dr. Ward of Indiana" who presented him with wooden sticks or tablets (possibly birch bark) upon which were drawn pictographs made by Delaware Indians. Each pictograph applied to a verse to be chanted, so that the drawing was a summary of many words and served as a mnemonic device to assist the chanter. Such representations were not unknown among the Indian nations, but none were written by northeastern tribes, although the Delawares were known to keep genealogical records.[1] Rafinesque stated that Dr. Ward had obtained the document as a gift in 1820 from Delaware Indians living on the White River in Indiana who were grateful for his medical care. Two years later, Rafinesque obtained, from an unknown source, another document that contained written song verses in the Delaware language associated with the pictographs on the painted sticks. Rafinesque claimed that at the time he did not know the language, either written or drawn, nor could he find anyone to translate them so that he was ignorant of their meaning and of their relationship to one another. It was only after his return to Philadelphia in 1826, where he had access to the work on the Delaware language by Moravian missionaries, David Zeisberger and John Heckewelder, that he was able to translate the material he had gathered in Kentucky. Besides documents on the Lenape language, he made extensive use of a manuscript dictionary in the library of the American Philosophical Society, written by Zeisberger, so that by 1833 the translation was completed. The Delaware words were associated with each pictograph accompanied by an English translation, all contained in two, forty-page notebooks. The original work was in five books, or songs, containing 183 verses in all.

The translation was first published in *The American Nations* (1836), along with twenty verses that extended the history of the Delaware Indians from about A.D. 1600, when the *Walam Olum* closes, to A.D. 1800, a time when the Delaware Indians were residing in Indiana. These later verses, which had been translated by a John Burns, were found on a "fragment" of unknown provenance and were undoubtedly authentic, according to the archaeologist C. A. Weslager.[2]

When Rafinesque died in 1840, his library, manuscripts, specimens, and some of his personal effects were sold to cover the expense of burial, but the bulk of his papers—cartloads—were assigned to the rubbish heap. It was an egregious act of vandalism and malice by the Philadelphia establishment, some of whom were happy to see the last of him. Oddly enough, one of his sternest detractors, S.S. Haldeman, purchased a few of the manuscripts on Indian mounds and the notebooks containing the *Walum Olum*. There is no record of his purchase of the original wooden sticks, so whatever their history or whether they even existed, they have been lost forever.

The documents that came into the possession of Brantz Mayer, a Baltimore lawyer and historian and the founder and first secretary of the Maryland Historical Society, were loaned to E.G. Squier, the author of *Ancient Monuments of the Mississippi Valley* (with E.H. Davis). Squier made extensive use of the Rafinesque material, translating the legends for a second time, and wrote a detailed commentary on the *Walam Olum*. Despite what he knew of Rafinesque ("it is usually safe to reject his conclusions"), and his suspicions, he concluded that the *Walam Olum* was an authentic Indian record and that Rafinesque's translation was "a faithful one." To Squier, not only were the pictographs an aid to memory, there was also adequate evidence that they were used by the Delawares to pass tribal legends from one generation to the next with some degree of accuracy and that they could be understood by many tribes.[3]

The first outspoken and vigorous skeptic of the documents was Henry R.

Schoolcraft, an authority on Native Americans, who felt that the *Walam Olum* glyphs were questionable because they seemed to resemble archaic Chinese figures. Peter Duponceau of Philadelphia, the foremost authority on the language of the Delaware Indians and one who considered Rafinesque a charlatan, simply ignored anything written by Rafinesque, despite the fact that the *Walam Olum* could be an important chapter in the history of humankind, which would complement Hebrew, Greek, and Chinese legends and, in fact, would corroborate Rafinesque's unitarian view of early human history.

In the early 1880s, Dr. Daniel G. Brinton of the University of Pennsylvania, a wealthy Philadelphian who was one of the foremost experts on Native Americans in the United States and was a founder of American anthropology, purchased the Rafinesque manuscripts from the Mayer estate. He was a fair-minded man, and while fully aware of Rafinesque's reputation, he seemed fascinated by him and was willing to devote years determining the authenticity of the tribal chronicle—mainly because of its enormous potential importance. The epic was, in essence, a Bible or Koran, a tale comparable to those great voyages and migrations of antiquity—an Iliad, or the wanderings of the Hebrews. It was a tale rich in events, heroes, and villains that has been included in studies of Native American history and literature as a grand, heroic poem, *Brotherly Love,* based on legend and some historical fact that provided insight into humanity's earliest moments.[4]

The essential veracity of the *Walam Olum* was defended by McCutcheon, who believed that the "vagueness and ambiguity" of existing translations prevented a full appreciation of the work. So he attempted a new translation to "clear up some of the confusion surrounding the interpretation of the Wallam Olum, and to turn it from an academic curio into a piece of living history." Whatever the merits of the translation, it does not address the question of authenticity. McCutcheon's passionate yearning to believe that the document is truly the noble record of an ancient epic, however beautiful and however useful its authentication might be, is no substitute for critical analysis of every kind. In a sense, this is Rafinesque revisited.[5]

With the help of scholars and educated, native Delaware speakers, Brinton made a new translation of the verses, which he included in a book that also contained the original Delaware words and glyphs copied from Rafinesque.[6] To identify the mysterious Dr. Ward, Brinton journeyed to Kentucky and Indiana, following every lead to track him down, but he met with no success, nor could anything be discovered about John Burns. There is some evidence—not incontrovertible—that the Dr. Ward in question was a Malthus A. Ward, a "botanical friend" of Rafinesque.[7] Although he considered it was "an insult to the memory of the man," Brinton felt compelled to ask whether the *Walam Olum* was a forgery, because even after fifty years there still lingered about Rafinesque a whiff of fakery and unreliability.

Detailed analysis of Rafinesque's notebooks, with the help of other experts, led Brinton to believe that the text was "of aboriginal origin," and that it was a "genuine *oral* composition of a Delaware Indian." The errors in the text, the incorrect meaning given to some words, and the obscurity of others not found in Zeisberger's dictionary, convinced Brinton that the document could not have been invented, and that the text and the figures were authentic pictographs of Algonquian origin. It was a "genuine native production," although not ancient, but worthy of study. "The narrator was probably one of the native chiefs or priests . . . with some knowledge of Christian instruction" but who "preferred the pagan rites, legends and myths of his ancestors."[8]

Bizarre similarities of the *Walam Olum* with the records of other tribes and cultures were explained away, and one senses that Brinton wanted to believe in Rafinesque and the documents. He gave Rafinesque credit for being the first to realize that "the Indian pictograph system was based on gesture speech"—essentially, sign language—an original idea that was particularly relevant at this time (the 1880s) because sign language was an active field of study to improve the lot of deaf-mutes by providing them with a means of communication. Brinton's seal of approval in 1885, however qualified, carried with it great influence, so that over the years the *Walam Olum* was taken up by many well-disposed scholars—but the revisionist movement was too late for poor Rafinesque.

C.A. Weslager (1972), who was convinced that the *Walam Olum* was authentic, discussed the possibility that Rafinesque might have borrowed the migration story that was already known by John Heckewelder and was consistent with Algonquian legend,[9] to which he had added the tale of the crossing of the Delawares from Asia across the Bering Strait, but this proposal was dismissed for lack of convincing evidence.[10] Voegelin (1940) examined the *Walam Olum,* a complex expression of a culture, for elements common to other tribes, because suspicion would be cast on the authenticity of the *Walam Olum* if it were utterly unique. Indeed, he found among other Delaware and eastern tribes antecedents' pictographs on painted sticks, a deluge story, and genealogical recordings. However, the combination of these components among the Lenape was unique—an acceptable conclusion. It would seem that information had been sought to support a position, and it was found.[11]

According to Newcombe (1955), there was little reason to doubt the authenticity of the *Walam Olum* as a document of the Delawares with possible infusions of other Indian mythologies. However, there was no linguistic or archaeological evidence to support the migration story related in the document—the southern and eastern migration account was found only in the *Walam Olum*. Newcomb suggested that the heroic tale was a late eighteenth-century production by a Delaware leader who created an epic mythology, one harkening back to a golden age, in order to revive the pride and sense of identity of a dispirited people—scattered, alcoholic, dispossessed of their land, and caught in the colonial wars between various European nations. But however useful the *Walam Olum* might be, it was inaccurate from an historical point of view.[12]

Despite all the investigations and theories, few were willing to give the *Walam Olum* an unqualified endorsement as a true, ancient history of the Delaware nation. In 1954, a team of about twelve investigators, including experts on Indian history, linguists, ethnologists, and archaeologists published an extensive report on the *Walam Olum*—a handsome tome. The concerted and meticulous study, which had lasted for twenty years, was sponsored throughout the Great Depression of the 1930s by the Indiana Historical Society and wealthy benefactor Eli Lilly, whose family owned a pharmaceutical business. Lilly, a dedicated amateur student of Native American culture, was deeply engaged in the study. Linguists made a new translation using the latest, highly technical linguistic methodology, and in consultation with existing Delaware-speaking Native Americans in Oklahoma and Ontario. A heroic but unsuccessful effort was made once again to identify the Dr. Ward who had given the original painted sticks to Rafinesque. Lilly himself attempted to date the events in the *Walam Olum*, place them geographically, and link them with other known historic occurrences. Physical anthropologists and archaeologists added to the picture; their results were "not always conclusive," nor were they encouraging. At best, all evidence was inferential. Although they were convinced that the *Walam Olum* was an authentic, meaningful record, they could not be sure, because "the veracity of the document in all its parts cannot be proved archaeologically, neither can it be disproved." Archaeologist Glenn A. Black concluded his report: "[T]he faith that some of us had in the document twenty years ago has not diminished. If it had, the tremendous amount of work by many individuals which has gone into this volume would never have been expended." The volume referred to was indeed a beautiful, lavishly produced publication. The team's position was defensive as it fended off the skeptics and it's own inner doubts, and it is difficult not to believe that during the hard years of the Great Depression, the team took its cues from their generous and enthusiastic leader, Eli Lilly.[13] There were always those who were convinced that Rafinesque's *Walam Olum* was a fraud, including an archaeologist formerly associated with the Lilly group. The uncertainty survived for 150 years for want of definitive information, and critical textual analysis.

In 1994, Oestreicher, an expert in Indian ethnography, published just such an analysis that led him to the conclusion that the *Walam Olum* authored by Rafinesque was unequivocally a fraud.[14] Oestreicher, who studied the Delaware Indians for many years, pointed out that recent carbon 14 dating had shown that there had been Lenape ancestors in the northeastern United States for 12,000 years, far more than the 3,600 years that Rafinesque had calculated from his document. He found "that elderly Lenape speakers did not consider the *Walam Olum* part of their culture at all. . . . [T]hey found its text puzzling and often incomprehensible." Examining the text, Oestreicher determined that it was "filled with preposterous grammatical constructions," and in places Lenape words had been crossed out and substituted by others. He concluded that a process of translating English verses into Lenape rather than the reverse had taken place. Rafinesque had composed the Lenape myths in English in accordance with his own version of their history, (crossing the Bering strait, and migrating south and east to New Jersey), and had then translated them inaccurately into the Lenape language. Rafinesque's Lenape words had come from Zeisberger's and Heckewelder's works, as well as from several other sources, but he did not have mastery of the grammar. A detailed, word-by-word analysis of the *Walam Olum* revealed that "the text was full of fractured constructions, misused words, including Lenape versions of English idioms . . . other Delaware words were contorted to resemble Hindi, Hebrew, and Greek." Oestreicher concluded that Rafinesque had a superficial understanding of the Delawarian language that led him to make mistakes—really howlers—that would not be made by a native speaker. As for the glyphs, they fared no better. They were "hybrids concocted from ancient Egyptian writing, ancient Ku-Wen script, Ojibwa Midewiwin pictographs, and even some Mayan symbols." Oestreicher claims that historical and typographical errors found in their work, and some non-Delaware words, were carried over into Rafinesque's document. Some of Rafinesque's Delawarian words were found to be of Swedish and Dutch derivation, proof that the version supposedly translated by Rafinesque could not be ancient, nor could it be earlier than the seventeenth century. Rafinesque's lists of chiefs in the *Walam Olum* were identical to those related by Heckewelder in his history of the Delawares of an earlier time. The names were reused by Rafinesque, and according to Oestreicher, the names of the chiefs of the Delawares were rarely, if ever, used twice. Further damning evidence was presented that Rafinesque had used an essential list of Delaware names already published by Heckenwelder, but Rafinesque had claimed that he had translated the *Walam Olum* before he had seen this list; Oestreicher shows that this could not be so. From Rafinesque's unpublished notes and drawings in the archives of the American Philosophical Society, from which he had written the *Walam Olum*, Oestreicher claims to have located the sources of most of Rafinesque's text, written and drawn. His argument is detailed and convincing, and his conclusions are compelling.

If the *Walam Olum* is indeed a fraud, what was Rafinesque's motivation in its perpetration? Rafinesque had stepped over a critical boundary—from bending the truth, proving a point by selection of data, and distorting facts to bald-faced lying and dishonesty. Rafinesque worked on the *Walam Olum* in the last part of his life at a time that reveals the aging and breaking down of a tired man, perhaps driven to the point of outright falsification by his frustration at being shunned after so many years of hard work. Perhaps he wanted to enjoy one final, dazzling triumph, proving his long-held theories correct, and seeking vindication for his years of ignominy. It would seem that in an age of discovery, Rafinesque desperately wanted credit for a major, unifying theory of global dimension that revealed one of the earliest migrations of humanity, the true origin of the Native American, the destruction of the early mound-building civilization in America, and the decoding of the earliest written Native American language, accomplishments that would rival Champollion's acclaim for deciphering hieroglyphic script. He felt that he deserved this fame and that with it would come financial support.

Weslager doubted that financial gain was a prime motivation, for there would be little profit in selling his translation of the *Walam Olum*, and in fact he was busily involved in other matters at the time of its publication in the *American Nations*. On the other hand, Oestreicher has argued that Rafinesque was indeed very concerned about financial matters; he proposed the *Walam Olum* for a prize of twelve hundred francs offered by the Royal Institute of France for an essay on the Delaware and other Indian languages, and he made several unsuccessful attempts to obtain a pension from King Louis Philippe of France for his great achievements.[15]

Perhaps the exercise was inspired by the well-known experience of Joseph Smith, founder of The Church of Jesus Christ of Latter-Day Saints, who in *The*

Book of Mormon had claimed discovery of tablets describing the descent of Native Americans from the Jews, a notion rigorously denounced by Rafinesque. A few years later he had his own *Walam Olum* version of the history of Native Americans that differed from Smith's.[16] Smith and Rafinesque covered some of the same ground in their legends, and in some details there were similarities. Smith had discovered his tablets under mysterious circumstances and they had inexplicably disappeared, as had Rafinesque's painted sticks.

Other elaborate linguistic frauds had been masterminded in the past so that the scholarly community was cognizant of this possibility. Brinton was fully aware of a hoax perpetrated by two French linguists who had created a grammar and songs of the Taensa tribe, living in the Mississippi Valley. He played a large part in exposing that fraud, and yet he did not seem to apply the same acuity to Rafinesque's opus.[17]

Hovering over this unfortunate tale is the possibility that Rafinesque himself was the victim of a hoax perpetrated by the true forger of the *Walam Olum*. Its author could have been the mysterious Dr. Ward, who has never been identified properly, or someone who gave the document to Dr. Ward. Rafinesque had displayed an uncritical enthusiasm and gullibility in publishing an account of Audubon's fish, and in a similar vein he may have been convinced that the Lenape epic he received was authentic. Would that this kinder explanation be true.

Chapter 13

BOTANY AND ZOOLOGY

The era between the publication of Linnaeus's *Systema Naturae* and Darwin's *Origin of Species,* two of the greatest landmarks in the history of biology, was one of identifying and classifying organisms, and the greatest of all identifiers of species at that time was Constantine Samuel Rafinesque. Although the breadth of his interests was remarkable, there is no doubt that he was primarily a botanist and a zoologist, especially concerned with rational order in biology. He was a natural historian rather than a scientist, cataloguing everything in sight—birds, molluscs, fish, snakes, lizards, turtles, insects, sponges, crustacea, and mammals[1]—creating systems and categories into which any living thing could fit. His curiosity extended to the nonliving—fossils, rocks, minerals, and Indian mounds; he even recorded and categorized weather and atmospheric conditions with some regularity, trying to discern rules for what appeared to be randomly changing states, with the hope of being able to predict the weather. However he described and classified (with little interest in probing beneath the surface) to determine how things worked, preferring to formulate profound (and pseudoprofound) rules of nature; it did not occur to him (nor to most of his colleagues) to harvest new information through experimentation. Usually, scientists had some background based upon formal education and training, but this Rafinesque lacked, as he did scholarly caution, and so he might be considered a talented, full-time amateur, learning as he went along. Extraordinarily well-read, he must have been aware of virtually everything that was going at the time in science and natural history.

Rafinesque's American and European contemporaries were excited about the fabled natural treasures of America, and they were eager to augment the great Inventory of Nature. However, Americans had first call for searching out the flora and fauna at their doorstep. Their descriptive and classifying approach was a popular and necessary first step in keeping with the systematizing philosophy of the Enlightenment and Utilitarianism, which melded with the wondrous, lyric creations of the Romantic ethos.

Rafinesque could hardly help being *the first* to describe so many forms of life in the American hinterland, which was largely unexplored. As Jordan wrote: "He was the frontiersman of our natural history, the Daniel Boone of American science."[2] Making the task of identification difficult was the virtual absence of publications on the flora and fauna of the country, and the few that did exist, such as Alexander Wilson's *American Ornithology,* were too bulky to be carried about in the field.

The name *Rafinesque* in taxonomy has had a curious history. In his day, dependent on unsympathetic peers who neglected his work, the name was disdained and it fought to survive. But for the past hundred years, as the Rule of Priority in naming organisms became a law of science that one could only thwart by ill-tempered casuistry, *Rafinesque* will not be denied without due explanation. Dispassionate analysis of the record shows that he was indeed the first to describe—however imperfectly—and name a universe of plants and animals. If the Rule of Priority holds for validly described and published plants and animals, Rafinesque's naming must be used, and the word *Rafinesque* must remain as the author of the name.

Beginning in 1814, while in Sicily, Rafinesque published four grand, encyclopedic outlines of the universe—ordering and categorizing all its organic, inorganic, and living components.[3] He divided all life into two kingdoms, ten classes in each, and in these were 2,500 genera of plants and two thousand of animals. Rafinesque redefined rules of taxonomy with the aim of "improving" upon the rules of his master, Linnaeus; at points, the two systems were incompatible, adding to the chaotic state of classification if both sets of rule were applied. The ideas and classifications presented in these publications were the foundation of all his future work, either as outright reiteration or, occasionally, elaboration, and there is little question that he borrowed from others. In Sicily, growing frustration and agony, both personal and professional, were explicit in his concluding remarks of the *Epitome* where he fulminated against the Sicilians, blaming them for any shortcomings in his work.[4] Rafinesque was never the man to moderate his opinions, and so the resentments and loathing heaped on him by the Sicilians were returned in kind.

But no matter, for Rafinesque's ambitious productions were ignored and had little impact. Emblematic of this neglect are the uncut pages of George Cuvier's copy of *Analysis of Nature,*[5] and an uncut copy of *Caratteri . . .* in a Dutch museum of Natural History, illustrating the lack of interest in reading

Rafinesque's lofty formulations and his notion of instability of living forms. The source of the ideas was suspect, and perhaps this sort of elaborate ordering in the manner of the eighteenth-century Encyclopedists was becoming tiresome and going out of style. By the time of Buffon's death in 1788, French natural historians had had enough of his grand cosmogonies.[6] In a society of investigators that was becoming more specialized and reductionist, where the Baconian ideal of fact-finding was becoming the rule, grand schemes and speculations, all-embracing inventories without new, hard data were losing their impact. Indeed, observation and description, not ambitious contrivance, have always been the major elements of natural history, and though experimentation and theorizing may have been placed in service, they were of secondary importance.[7] A few decades after Rafinesque's death, Darwin's theory was taken seriously because it was supported by vast amounts of data and finely reasoned argument that could explain so much.

Botany

Rafinesque, "the most erratic botanist of all time," declared many times that botany was "one of the most useful and amiable sciences." From the age of twelve he collected and preserved plants, and in the course of his career he formed several huge herbaria, which grew rapidly as he scoured the countryside.[8] His collection also grew through gifts and exchanges with colleagues in all parts of the world so that his herbarium was global in origin, and his pronouncements on the classification of his specimens had repercussions in both Europe and America.

In theory and in practice, Rafinesque stressed the value of the herbarium in botany, but it was his luckless fate to have his collections destroyed or dispersed like no other botanist, before or since. These lost collections contained Rafinesque's type specimens by which botanists could verify his claims. Alas, since very few survived the depredations of time and neglect, his frail reputation for reliability suffered yet another crippling blow.[9]

Acknowledged by some experts as the foremost authority on plants growing beyond the Allegheny Mountains, he was probably the first in America to write about ecological succession of plants in a given area: "There is therefore a kind of natural perennial change of vegetation; when a species has exhausted the soil of a peculiar nutrition that it requires, it gives way to another for a series of years."[10] He was an advocate of land management—abusive practice had led to the disappearance of cane grass upon which cattle fed, especially in winter, and so he recommended special cultivation of grass for harvesting in winter. He also recommended the cultivation of wild rice, silkweed, and sugar maple trees, which could supply the nation with sugar.[11]

As a professor at Transylvania University, his recorded lectures on botany revealed a mature admiration for his subject, lauding its endless variety, its inherent interest, utility, and beauty. He wanted his students to experience the sublime joy of studying plants as he had, and from the recorded reminiscences of former students, many of them young women, he seems to have been successful. Rafinesque insisted that science and natural history should have an important place in peoples' lives: "We can hardly find an object [a flower or plant] more deserving to employ our leisure. . . . A single blade of grass, or blossoming plant, will afford us an unremitted enjoyment and might afford us occupation during many years if we were to consider it and study it under all its different points of view." Nor did Rafinesque neglect the practical aspects of science and botany, for one of Rafinesque's messages to the students was that the future success of their country depended on the practical lessons learned from botany and science.[12]

The most sympathetic of Rafinesque's admirers, E.D. Merrill and M. L. Fernald, had to admit that Rafinesque had made "unending trouble" for American and European botanists, and for the same reasons, he had also thrown other fields as disparate as history, ethnology, linguistics, ichthyology, conchology, and ornithology into a similar disarray. But of all these disciplines, botany has been the most closely examined by his colleagues. He had contributed to each, but he had changed rules quixotically and imposed his own slack boundaries of genera and species that differed from all others.[13]

By his own rules he had proposed 6,700 species and nine hundred varietals, 2,700 genera, and 320 subgenera, but only thirty were generally accepted, a remarkably low rate. Merrill has calculated that less than 5 percent of his names could be used strictly on the basis of priority, while 99 percent of Linnaeus's names have been adopted. In 1949, more than one hundred years after Rafinesque's death, 740 of his validly published generic names and 2,600 specific names were not even listed in standard botanical indices.[14]

This appalling neglect is due to the fact that Rafinesque was simply wrong in so much of his work and in direct conflict with all reasonable botanists, and so they felt his neglect was not a terrible injustice. Rafinesque was clearly in error in his identification and naming of species of *Clintonia, Dodecatheon* and *Trillium.*[15] Of the 1,210 taxonomic entities, and 654 species of Indiana plants that Rafinesque identified, only thirty-three of his names are now recognized—a poor record indeed![16] A century and a half ago, Asa Gray dismissed Rafinesque's work stating that of sixty-six new genera proposed by him, fifty were "absolute nonsense," and he ridiculed his claim that there were thirty-three indigenous species of North American rose.[17] Merrill, a critic who has rendered so great a service to Rafinesque, concluded that any botanist who wanted to reinstate a Rafinesque name for one already in use, thereby reducing it to synonymy, was in for trouble. He or she would be resisted not only by the author of the name, but by the International Botanical Congress as well, where such matters are finally decided. Most of Rafinesque's names, even after being removed from the "overlooked" category, will remain *ignotae vel incertae*—unused and nowhere—an unsatisfactory category."[18]

Since Rafinesque's time, the unearthing of his "overlooked" species and genera has become a minor industry. Pennell reported eighty-three, while Merrill published a list of 744, and he estimated that there may be as many as three thousand names to be uncovered—flowering plants, ferns, algae, fungi, and lichens.[19] Rafinesque's genera and species of plants listed in Merrill's *Index Rafinesquianus* is a complete accounting, and most of these are validly published, but only a few of the names have been reinstated—from being "overlooked" or from synonymy. Britton pointed out that Rafinesque was the first to distinguish the hickory from the walnut tree, and that he was the first to give it a proper name *Hicoria* in 1808 (*Hicoria,* Raf., 1808), but the name was ignored by Thomas Nuttall who renamed it *Carya* in 1817 (Nuttall, 1817), with no explanation. Britton insisted that Rafinesque's name should hold.[20] Here and there, over the years, a Rafinesque name has come into usage through the insistence of a botanist or zoologist who has read the record, but on the other hand, the botanical literature is filled with papers arguing why particular Rafinesquian (nonconserved) names are erroneous, invalid, or have been anticipated by others and therefore should not be used. According to the eighth edition of Gray's *Manual of Botany,* from a multitude of Rafinesque's taxa, only eighteen genera, eighty-four species, and thirteen varieties of Rafinesque have been fully recognized—a final tally as of 1950.[21]

Wherever he had been, he studied the botany of the region assiduously—in Sicily, Kentucky, the northeastern United States, and in the Ohio River Valley, always striving to capture as large an audience as possible, often by writing out the name of a plant in Latin, English, French, and German, and giving the pharmaceutical and the common, vulgar names by which a plant was known.[22]

Fish

On his first voyage to America as a mere nineteen year old, Rafinesque had described and drawn a few fish and molluscs, but his first major foray into this world occurred when he resided in Sicily from 1805 to 1815. As attested to by his colleague, William Swainson, he examined fish that were either alive or freshly caught, thus avoiding the errors attendant on studying discolored, preserved specimens.[23] Most of this work appeared in his first two publications, therein are described 390 species of fish, of which 190 were new,[24] forcing even a tireless critic like S.S. Haldeman to admit that the work was "good."[25] Swainson was impressed by this pioneering work and was prompted to comment that not infrequently, Rafinesque anticipated the renowned Cuvier and Valenciennes but did not get credit for his findings. The reputations of these famous investigators and their words were so unassailable, that when they treated Rafinesque with contempt, they were not challenged, and Rafinesque's work was dismissed and considered unworthy of citation. New names were assigned because of Rafinesque's allegedly poor descriptions or because Rafinesque described fish from geographic regions other than those studied by Cuvier and Valenciennes, and were, so they reasoned, different species.

Following the publication of nine specialized papers in his *Western Review and Miscellaneous Magazine* identifying new kinds of North American fish,[26] Rafinesque brought out his major zoological work *Ichthyologia Ohiensis,* which described for the very first time the fish of the Ohio River—111 species—and the work also included a physical description of the river and its tributaries. The book's appearance was hardly acknowledged, and its contents were completely excluded from Albert Gunther's important and widely consulted *Catalogue of Fishes.* Aside from the imaginary fish created by Audubon and included in the monograph by Rafinesque,[27] the work was praised decades later as original and valuable, one of Rafinesque's best efforts, and as reliable and worthwhile as anything done at the time. Audubon's fish were a confounding element in an otherwise excellent work.

Rehabilitation of Rafinesque's reputation in the field of ichthyology began in earnest with Kirtland's studies of the 1840s, and Louis Agassiz's authoritative article in 1854 on fishes of the Tennessee River, in which he had kind words to say about Rafinesque's work.[28] Discussing the taxonomy of Cuvier and Rafinesque, comparisons were implicit, and it is fair to say that Rafinesque held his own against Cuvier's high-handed neglect, noting that Cuvier had renamed genera first described and named by Rafinesque. Agassiz had no difficulty identifying fish using Rafinesque's descriptions, stating "I do not hesitate in giving him the fullest credit. . . . Both in Europe and America he has anticipated most of his contemporaries in the discovery of new genera and species in those departments of science he cultivated most perseveringly, and it is but justice to restore them to him [genera and species] whenever it can be done." In 1844 he wrote his colleague, Charles Lucien Bonaparte, "I think there is a justice due to Rafinesque. However poor his descriptions, he first recognized the necessity of multiplying genera in ichthyology, and this at a time when the thing was far more difficult than now."[29]

One of the lessons learned from Agassiz's paper was that *everyone* was confused at times and fell into taxonomic *error*—even the illustrious Cuvier—but when Rafinesque made mistakes, he was branded *unreliable*, which was justification for ignoring his work and priority. However, Agassiz did criticize the procedures of both Rafinesque *and his contemporaries*: "Nothing is more to be regretted for the progress of natural science in this country than that Rafinesque did not put up somewhere a collection of all the genera and species he had established with well-authenticated labels, or that his contemporaries did not follow in his steps, or at least preserve the traditions of his doings, instead of decrying him and appealing to foreign authority against him." Agassiz, a confident cheerleader of American science whose credentials were beyond challenge, was reflecting a sentiment articulated by Emerson and shared by many American scientists that they should no longer defer abjectly to European dominance.

Agassiz's comments were truly the powerful opening gun for the move to rescue Rafinesque's name and reputation, (and to instill in American scientists confidence in their own abilities, vis-à-vis Europe). Rehabilitation continued in Charles Girard's comments about Rafinesque so that he felt it highly desirable to authenticate and restore Rafinesque's genera and species. This could best be done by retracing Rafinesque's footsteps along the Ohio River and its branches with Rafinesque's texts in hand, and this is, indeed, what David Starr Jordan did to become a major actor in the rescue.[30] However, Copeland's *A Neglected Naturalist*, unalloyed praise for Rafinesque, and published in the broadly read *American Naturalist* (1876), was also a great boost to his reputation.[31]

In 1877 Jordan, a leading ichthyologist (who became the first president of Stanford University), published a careful, definitive review of the North American fish described mainly in Rafinesque's *Ichthyologia Ohiensis*.[32] Jordon did not have complete confidence in the data, because there were no type specimens and he was well aware that some of Rafinesque's descriptions were not done at the time of examination during the summer, and that Rafinesque did not measure with a ruler but *estimated*. Nevertheless, taking these negative factors into consideration, Jordan concluded: "I have succeeded in identifying more or less satisfactorily, nearly all of his species, and in restoring to a number of his names their rightful priority." In essence, Jordan reexamined the fish of the rivers described by Rafinesque, and wrote a "corrected" and annotated version of his book. For the most part, the original version of *Ichthyologia Ohiensis* was correct and of value, but it could not be accepted in toto but simply inserted into the existing record, because it contained some errors of description, serious misprints, and confusing points in nomenclature. In all, "of the 79 genera and 115 species of fishes known as inhabiting the Ohio and its tributaries, 29 genera and 37 species were first described" by Rafinesque.[33] In this instance, many of Rafinesque's names were restored through the efforts of two giants in the field, Agassiz and Jordan. In his biography of Rafinesque, Call, an ichthyologist, states that Rafinesque had extraordinary "diagnostic talents," and that *Ichthyologia Ohiensis* is the "groundwork of the ichthyological literature of the great valley of the Mississippi River."[34]

Since Rafinesque's curiosity was boundless, he not infrequently was attracted to bizarre subjects, which he treated seriously and respectfully. One such study was entitled *Dissertation on Water Snakes, Sea Snakes and Sea Serpents,* which was published in more than one journal.[35] He carried out his study to sweep away confusion from the mind of the public and to expose the "speculative writers," whose wild conjectures embarrassed serious American scholars and scientists in the eyes of Europeans. Rafinesque was well aware of the precarious reputation of American science in Europe. What followed was a learned treatise, which distinguished eels, a true fish, from snakes that live part of their lives in water; Rafinesque identified and classified numerous examples of each. From this careful account, he continued in the same tone to describe such species as the Massachusetts sea serpent, the Lake Erie sea serpent (thirty-five feet long), the scarlet sea-serpent (forty feet long), and Captain Bowen's sea serpent, whose head and neck stood up like a mast, all reported by sailors and others in the near and distant past. Rafinesque wrote about the Massachusetts sea serpent:

> From the various and contradictory accounts given of this monster by witnesses, the following description may be collected—It is about 100 feet long, the body is round, and nearly two feet in diameter, of a dark brown, and covered with long scales in transverse rows; its head is scaly, brown mixed with white, of the size of a horse's and nearly the shape of a dog's; the mouth is large, with teeth like a shark; its tail is compressed, obtuse and shaped like

> an oar...It is evidently a real Sea Snake, belonging probably to the genus *Pelamis,* and I propose to call it *Pelamis megophias....*

Other monsters were described with the same solemnity.

For Rafinesque there was more natural history here than there was folklore. Just as he was taken in by Audubon's fantastic fish, he seems to have believed the accounts of these monsters, classifying the giants, and giving them a binomen. American naturalists, ever commonsensical, dismissed these tales for the nonsense they were.

Molluscs

The first work on molluscs of the United States in the early decades of the nineteenth century was carried out by C.A. Lesueur, a native of France who spent several years in Philadelphia, Thomas Say, perhaps the best of all early American naturalists,[36] and Rafinesque.[37] Alas, Rafinesque's fate in conchology was the same as in other fields—relentless neglect and modest reinstatement. In conchology however, there was a particularly fierce, and mean-spirited, grasping for priority; his rejection was nothing less than a conspiracy. Members of the establishment—closet naturalists of great influence—chose to dismiss his work, justifying their actions by claiming that they could not make heads or tails out of his papers. But later, dispassionate critics, reviewing the same publications, came to different conclusions and correctly pointed out Rafinesque's priority in more than a few instances.

Soon after Say had published his work on new species of North American freshwater bivalve molluscs—the unionids—Rafinesque described many other new species and genera and five new subfamilies, which he discovered while traveling from Philadelphia to Kentucky. He was the first to study the numerous unionids of the Ohio River, many of which he sold or sent to his European colleagues.[38] The waters he investigated were so remarkably rich in different forms of molluscs that his contemporaries, unfamiliar with the locale, did not believe him. But Rafinesque recognized that his specimens were different and in vastly greater variety than the European *Unio,* in which there were only a few genera. This was a first important conclusion in his major work on North American unionids, which, according to one authority, was twenty years ahead of its time.[39] Rafinesque also introduced the term *malacology,* but this was superseded by *conchology*.

In 1819, Rafinesque described and named the Ohio River mollusc,

Campeloma crassula, an original finding that was ignored by American conchologists until 1864.[40] The genus *Campeloma* was described and given other names after 1819 by several naturalists including Isaac Lea, who wrote an exhaustive treatise on the molluscan family *Unionidae* (freshwater mussels), and the genus *Unio*. Before Rafinesque, American naturalists had lumped all unionids into only four genera, but in fact there were almost 1,500 members with much variability, making it exceedingly difficult to determine which were species and which were varieties.[41] Despite Rafinesque's contribution, Lea almost completely neglected his findings—his assigned names and his system of relationships. Lea thundered that his work was "useless, incomprehensible" and "not deemed [sufficiently] important to insert here." To justify his virtual exclusion of Rafinesque, Lea also included several pages of censorious judgments by other naturalists regarding his reliability.

This rough treatment meted out to Rafinesque offended the conchologist, Timothy Abbot Conrad, who came to his defense, contesting Lea's criticisms point by point.[42] In answer to one complaint that Rafinesque's descriptions were inadequate and therefore his identifications must be discounted, Conrad claimed that they were no less detailed and indefinite than those of the revered Linnaeus and Lamarck, and so they should all "share the same fate." There was no difficulty defending the charge that Rafinesque did not cite previous work, for conchologists were forever complaining that others had not given them due credit—all were culpable. Even Lea had felt neglected by another conchologist, William G. Binney—there were fearful recriminations in every direction under a glaze of polite bonhomie. American conchologists, who gave new names to molluscs previously named by Rafinesque, were obviously ignoring Rafinesque's prior claims, just as they asserted he was ignoring theirs; Rafinesque did not remain silent about this neglect, but to little effect.[43]

For one reason or another, virtually all of Rafinesque's names were considered invalid. G.W. Tryon, who wrote a short history of the field, and William G. Binney, the author of a definitive treatise on terrestrial molluscs of the United States, heaped abuse on Rafinesque, although they admitted that though his early work in Sicily was commendable, it was his work in the United States that they reviled. Binney wrote that Rafinesque was "not an American, but for a quarter of a century a resident in the United States,"[44] and Lea considered Rafinesque "an intruder in this country,"[45]—the prejudiced views of members of the Philadelphia social and scientific establishment. One can only imagine how they accepted Rafinesque's boast, "I considered myself endowed with a sagacity for the perception of generic and specific differences far in advance of any man of my time," and what made it worse was the fact that he was largely correct. His reputation was further damaged by his claim that he was capable of establishing a pearl culture industry using North American mussels and would freely advise any adventurous soul willing to undertake the project.

Lea, a wealthy man and a gentle-looking scholar, was president of the Academy of Natural Sciences of Philadelphia and was perhaps the most influential scientist in Philadelphia. The very model of a detached, impartial Nestor, he zealously sought to expunge the name Rafinesque from the American record, and he could not contain his impatience with Rafinesque's defenders—both Conrad and Agassiz were objects of his displeasure. Lea was not happy with Agassiz's judgment: "If the American naturalists had followed Rafinesque's track instead of despising him, we should have gained a good while ago a treasure of important additions to the anatomy of the family." In reference to work on fish, Jordan ascribed Rafinesque's rejection to "conservative odium [that] always attaches to a writer who attempts to form natural genera out of time-honored artificial combinations.[46]

Rafinesque's larger vision of the problem of taxonomic categories was not appreciated by Lea and his companions, who were consumed with precise detail and an idealized, static view of life, at a loss to deal with variation. Despite the admitted shortcomings of Rafinesque's practice, he thought about the relationships between the different forms and how they could be derived one from the other by small changes over time using a scale that was greatly compressed compared to the established scale of the present day. In a period when the vastness of geologic ages was not fully appreciated, he estimated, in his *Flora Telluriana,* that it took "an average of 30 to 100 years for the deviating or splitting range of specific deviation [of plants], and 500 to 1000 years for the Generic deviation; altho' their real permanence is much longer," a time frame that seemed to derive more from the Bible than from Lyell and contemporary geology. His dynamic approach to speciation based upon constantly arising variations in form was born of vast field experience and imaginative thinking, especially in his consideration of plants. Although decades in advance of Darwin, Rafinesque's dazzling insight lacked Darwin's and Wallace's critically important idea of natural selection as the means of facilitating the evolutionary process. The notion of Rafinesque's "perpetual mutability," a dynamic instability, was unacceptable to an establishment that revered the notion of the fixity of species created by a God who would not permit the extinction of His own creations. Perhaps it was feared that the ongoing revision of the classification system of constantly changing living forms (especially of a genus like *Unio*) would lead to chaos.[47]

Because there was so much variability in molluscan forms studied by early American conchologists, Lea, the quintessential closet naturalist was particularly displeased by the species splitters who created confusion by introducing many new species of dubious validity. He was both policeman and judge, particularly motivated by the desire to show that American natural science was sound and dependable. He complained that "many European writers find it difficult to give us credit for our real disposition of keeping down our species as much as possible," and he was particularly unhappy when a French scientist claimed that not more than three in one hundred of Lea's species would stand the test of time, a wounding remark Lea considered "too flippant to need much notice." Still, it signified the anxiety of American naturalists and scientists to be accepted by the great arbiters of quality and originality—the Europeans.

While Lea and his contemporaries felt justified in excluding Rafinesque's work as incomprehensible, it is astonishing that in 1895, Call and others looking at the same accounts found that many of them were quite adequate for identifying Rafinesque's genera and species. In fact, Call states that his work was "remarkably well done," and that his descriptions were usually good. Rafinesque was the "first among many . . . [who] succeeded in regrouping, in some rational and natural manner, forms of most divergent character that before had constituted heterogeneous assemblages." Rafinesque's effort was historically important, the first who "called attention to the great wealth in the western waters of animal life." To Call, the story was an "unpleasant episode," and he could only draw the conclusion that several conchologists made a "concerted attempt . . . to ignore his work, and to reflect on his scientific reputation." More specifically he accused Lea of "not being entirely free of prejudice . . . bent to disregard Rafinesque's work."[48] By the early twentieth century, many of Rafinesque's names applied to molluscs were recognized. In a sober, detailed monograph of the genera listed in the family *Unionidae*, Rafinesque's name was attached to eleven out of eighteen on the basis of his descriptions, Lea's name was attached only to one.[49]

Perhaps it should come as no surprise that to a large degree, the acceptance or rejection of taxonomic work, which greatly depends on individual judgment, is subject to many factors, especially when there is ground for legitimate criticism. Without belittling objective criteria in science as in the arts, how the critic balances the good and the bad can depend on the personal and emotional regard of the critic for the author, the sense of confidence that the critic has in himself and in the author being judged at the time of decision, and the status of both the author and the critic (nationality, reputation, seniority). They were important determinants then, and they remain so now.

Birds

As the young Rafinesque sailed for the New World in 1802, he expressed the desire to write a book on the birds of America. So many of them were unknown, and those that were known, were "badly described." On his first stay in the United States he published on only one American bird, a canvasback, and he also carefully described four birds from Java that he found in the Peale Museum in Philadelphia. Two papers were sent to his friend, Daudin, who had them published in the *Bulletin des Sciences par la Société Philomathique, Paris.*[50]

Identification of birds began in earnest while he was in Sicily, where he described seventeen new species, and few, if any, are considered as valid today because he did not provide adequate information. In fact, one bird that he described in the Ohio River Valley, the citron open bill, has never been identified.[51] This bird may fall into the category of hearsay species that Rafinesque accepted at face value but never actually saw, as are four other birds, among which was the scarlet-headed swallow described to him by Audubon (along with his mythical fish).

Rafinesque's ornithological writings included a major classification based on the Natural System, in which he divided birds into one class, one subclass, six orders, twenty-five families, and about seventy-five subfamilies. This work was published in his *Analysis of Nature* and other obscure journals. A remarkable 181 new genera were created, but most were so inadequately described, really mere names and nothing else, that they were declared *nomina nuda* and were discounted. Some birds Rafinesque renamed because their original names were too long or too short or did not conform to his standards. Still, commenting on this classification, Richmond claims that if this work had been published in English rather than in French, "Rafinesque's names would be much better known than they are at the present day [1909], and our nomenclature would bristle with Rafinesquian genera, since he showed a discrimination of generic groups far in advance of his time."[52]

Rafinesque could write delightfully, as he did in an article in the *Kentucky Gazette* in which he described several birds of Kentucky, and in a second article, he wrote about the wandering sea birds of the Western States.[53] He could be generous in his praise of work that he respected. For instance, he lauded Alexander Wilson's *American Ornithology* as a "magnificent work. . . . it stands as a monument of genius," and he was proud of the fact that it had been published in "our country." Indeed, this early work was of particular importance in the history of American science because it was the first major American work to be recognized as a masterpiece by both Europeans and his own countrymen. At a time when there was uncertainty about American creative talent, Wilson's success heartened the nation, aroused great interest in ornithology, and prompted European naturalists to work in the United States.[54] Still, classification of birds was unsatisfactory, for there was no insightful, theoretical basis for their arrangement. In some instances birds were profusely and beautifully illustrated (works by Thomas Nuttall, Lucien Bonaparte, Titian Peale), but they were simply listed in the order in which they were found, clearly a failure in classification due to a lack of understanding of taxonomic principles, which lessened the scientific importance and usefulness of their efforts. Tensions arose when scientists, and especially closet workers, began to realize that it was necessary to dissociate the field-worker's sense of beauty and poetry of observed objects from their scientific status. Audubon's magnificent creations were much admired as art but to the scientist they were false rearrangements of nature for art's sake— "miserable fabrications," and "a clumsy lie."[55]

Rafinesque claimed that he had "ascertained and distinguished above 660 species" of birds, sixty of which had been described and named by him, but the work remained unpublished. Rafinesque's obscure essays were uncovered by Samuel N. Rhoads, an ornithologist, who published and annotated the text: "The White-fronted or Blue Bank Swallow of authors, is destined to go down into the history of nomenclature as a distinguished bird. It made so many narrow escapes of being improperly named in a binomial sense that it seems a bit humiliating for it to now be snatched from the laurel crown of Thomas Say and transferred, by the rights of priority, to a man who he undoubtedly despised and certainly ignored. Say was one of the coterie of Philadelphia naturalists that eventually drove Rafinesque and his contributions from any recognition by the Academy of Natural Sciences. Whatever Say may have lost, Rafinesque certainly gained in having won in the priority game of naming and properly describing the cliff, or eave or republican swallow as *Hirundo albifrons.*" Once again, writing a short article in the obscure journal the *Kentucky Gazette,*[56] Rafinesque stole the thunder from his peers' detailed, properly published treatises. Rhoads, a man with a wry sense of humor, concluded that it was a blessing that Rafinesque had not devoted much time to ornithology—peace at any price.

Carcinology

No creatures were unworthy of Rafinesque's attention and time, and all aroused in him an urge to classify them and to write a definitive treatise on the group to which they belonged. He never wrote *the* treatise on carcinology, but he did publish eight papers on the subject that have been carefully reviewed by Holthuis, who, while rightfully restoring some of Rafinesque's names on the basis of priority, pointed out how his practice led to a confusion that later had to be undone.[57] He considered Rafinesque's classification and nomenclature of the crustacea, published in his *Analysis of Nature* (1815), as "queer." Rafinesque in his various publications would use several spellings for the same name—the class *Plaxolia* was also spelled *Plassolia* and *Plaxomia,* the name *Crustacea* was at times also *Crostacei* and *Crustacés,* and he called the field of Carcinology itself *Plaxology, Plassologia,* and *Plaxologie.* Perhaps his own phenomenal memory failed him at times.

Holthius listed several examples of the rightful replacement of accepted generic and specific names by earlier, valid Rafinesquian names that had been "overlooked." Although he recommended adoption of some of these, there were so many other validly published names assigned by Rafinesque that if all the changes took place in too short a time, "they would greatly upset the stability of carcinological nomenclature," and "cause nomenclatorial confusion." Accordingly, as a reasonable and fair outcome, he thought that many, but not all, of Rafinesque's names should be ignored, and some old, familiar names retained despite the violation of the rule of priority, so only three of Rafinesque's nineteen genera, and two of forty-two specific names are actually in use. This compromise, perhaps not satisfying to anyone, reflects the confusion and uncertainties engendered by Rafinesque in yet another field.

Chapter 14

LAST YEARS IN PHILADELPHIA

His years of teaching at Transylvania University at a hasty end, Philadelphia became Rafinesque's headquarters in September 1826 and remained so for the last fourteen years of his life. He returned to a city that was no longer preeminent in the artistic and cultural life of the nation, nor was it any longer first in foreign trade, having ceded the honor to New York, Boston, and Baltimore. To its rescue came iron and a nearby source of power—anthracite—that was mined just up the Schuylkill River, so by 1828 Philadelphia had become a city of iron foundries and the leading manufacturing center of the United States. By 1830 a quarter of the nation's steel industry was located in the city that fabricated most of the nation's steam engines and almost half of its locomotives. All this was attended by the birth and development of the American labor movement in Philadelphia's wool and cotton mills, which employed 35 percent of the city's workers, making it by far the leading textile and carpet manufacturing center of the United States. During the 1820s and 1830s, major general strikes by textile workers (men, women, and children) and carpenters were mounted to reduce the working day from fourteen to ten hours, and the labor movement began publishing its own newspaper, the *Mechanics Free Press.* With Nicholas Biddle at the helm of the railroad companies and the Second Bank of the United States, Philadelphia became a major financial center, and its growing wealth was invested in a construction boom that made heavy use of marble in its buildings rather than traditional red brick. The Philadelphia of Rafinesque's time was caught up in an expanding American Industrial Revolution, belching smoke in some areas of the city, while in others it was still a peaceful and beautiful metropolis and a port with considerable eighteenth-century charm remaining—orderly, clean, and abundantly endowed with greenery and fine architecture.[1]

The year Rafinesque settled in Philadelphia, his brother Anthony Augustus died in Le Havre, a loss that did not seem to arouse strong feelings in him. Five years later his mother, who had wanted Constantine to visit her and perhaps settle in France, died of a stroke in Bordeaux. Rafinesque seriously considered relocation to Europe as an excellent opportunity to exploit his "discoveries," sell his collections, and publish his work, something he was unable to do in the United States because he believed he had too many "enemies" and too few funds. Only after a decade had Rafinesque made enough money in banking, in the sale of specimens, and in medical consultations to be able to print some of his work. Indeed, virtually every penny he had ever earned was spent either on publication or on his ramblings through the countryside. Perhaps he would be able to repeat Audubon's experience. Audubon was unable to get his work on birds published in America but met with great success in printing and selling his masterworks in Europe. At the time, the United States was far from the center of action, and success in the provinces was unheralded in the great European centers of learning. Audubon's triumph in France, Germany, and England was a major achievement of which Americans could be proud—something to be emulated by aspiring naturalists.

However much he desired to see his mother, he decided to stay in America, because there was too much political instability in Europe, especially in the Paris of 1830. Despite all his difficulties, "I prefer the calm and security of this country, improving every year by wise institutions and entire freedoms of action of travel; where so much is to be done and explored by science."[2] Rafinesque, whose manner was "foreign," and whose French accent was said to be almost impenetrable, formalized his allegiance to the United States in 1832 by becoming a naturalized citizen, basking in the sunny advantages of living in the young Republic.

His experiences were recounted in his *Life of Travels,* written in French in 1833 and later translated into English, with an addendum recounting the events of 1834 and 1835, and the whole work was published in 1836. The errors in Rafinesque's French suggest that after three decades he had somewhat forgotten his native language, but the English version was not free of spelling errors and other peculiarities. For so articulate a man, the account was merely a ramble through his life, surprisingly spare, turgid, and lacking in color. It can hardly be called an autobiography. It was more a catalogue of his activities, but was too brief to be useful or gripping, considering much of his turbulent life had been omitted. Still, as a log of travel, a litany of places visited and persons met, *Life of Travels* allows one to follow his trail, and to some extent it strings together in sequence his various publications and undertakings. Rafinesque had kept diaries since childhood, which were probably lost in the shipwreck off Long Island, and so his account was drawn from memory, which remarkable as it was, still deprived his adventure of rich, stimulating detail. Thirteen notebooks of his later years still exist, but despite the many particulars contained in them, Rafinesque saw fit to simply ignore this information in the final version of his autobiography.[3] An early biographer, Richard Ellsworth Call, suggested that the biography was little more than a long letter written in the first person singular to Rafinesque's sister Georgette, and not an exhaustive record. However, there is good reason to believe that Rafinesque desired to have the work published in the *Bulletin de la Société de Géographie*, although this never came to pass.[4]

During the winter of 1826–1827 Rafinesque sustained himself by teaching a course on "natural history on the earth and mankind" at the Franklin Institute, and in the spring he taught geography and drawing in a high school associated with the Institute. Classes were so large and demanding, he had little time for his other activities—writing and displaying his many specimens. One of his students left a vivid description of Rafinesque and his class. He was "a corpulent man, with queer French accent" who "sometimes became very angry with the class" when he lectured:

> His odd manner and dress attracted the boys who laughed and made fun of them, and his lot seemed not to have been an ideal one. Rafinesque was very large about the waist and wore wide Dutch pantaloons of a peculiar pattern, and never wore suspenders. As he proceeded with a lecture, and warmed up to a subject, he became excited, threw off his coat, his vest worked up to make room for the surging bulk of flesh and the white shirt which sought an escape, and heedless alike of his personal appearance and the amusement he furnished—was oblivious to everything but his subject." Although the image was not fetching, the student considered his teacher "a very able man.[5]

Rafinesque's teaching career in Philadelphia lasted two semesters—liberation arrived in late spring of 1827, freeing him for travel.

Eager to roam, he was brought by coach to the New Jersey shore, where he carried on research for several days; from there he went by steamboat to New York City, then headed north to Troy where he lectured, and passed the time with his good friend Eaton. Two days of travel by coach brought him to Boston, and the next day he attended commencement exercises at Harvard University where he buried himself in libraries. Near Boston, he stopped at Worcester to visit Isaiah Thomas of the Antiquarian Society who was so important to Rafinesque for publication of his work on Indian ethnography and languages. After crossing beautiful, hilly country, examining the flora close hand, he visited a Shaker colony in New Lebanon in New York State, which provided him with herbs for his patent medicine venture. Rafinesque then traveled down the Hudson River by steamboat to New York, where he "rested a while," and by September he was back in Philadelphia, although in such poor health that he did not resume his arduous teaching at the Franklin Institute High School. He confessed that in his weakened state, preparing for classes and controlling unruly students was just too taxing, but he began his medical career and patent medicine business, inspired by his own successful self-treatment of what he diagnosed as consumption—"fatal Phthisis"—brought on by "disappointments, fatigues, and the unsteady climate." Though utterly unlicensed, he began to practice medicine, as a specialist no less, in pulmonary diseases, calling himself a *pulmonist* just as specialists in dental matters called themselves *dentists*. He was America's first and probably its only pulmonist.

In the next few years, Rafinesque published his two-volume work, *Medical Flora of the United States,* a useful, scholarly work that was also a promotion of his healing remedy *Pulmel,* which of all his writings had the broadest appeal, and was the most widely read. After arranging for the manufacture of his cure, he traveled not only to satisfy his wanderlust and scientific interests but to establish distribution centers for his concoction. Dispensaries were set up in towns around Philadelphia (Easton, New Hope, Mauch Chunk [Jim Thorpe]), and locations in New Jersey, Staten Island, and Long Island. In 1832, *Pulmel* was in the hands of agents in twenty towns and cities, and in Paris, Rafinesque's sister Georgette and her husband, Paul Lanthois, were his representatives. Clearly, he was a forceful and effective entrepreneur.

Whatever Rafinesque's activities, none provided him with a financial cushion. Departments of science were being established in colleges, old and new, but as soon as a position opened it was offered to "young, unskilled, unexperienced or incompetent candidates" rather than older candidates like himself, and he had little chance against these younger men. Numerous states were establishing geological surveys to exploit their natural resources, but Rafinesque held out little hope of being hired. Constantly frustrated, he turned to a literary career for sustenance—surely a precarious choice toward self-sufficiency—and from his pen came the works of poetry *The Universe and the Stars, The World or Instability, Despondency,* and several scientific essays that were embellished with his poetic laments.

Some idea of his financial status can be gleaned from his *Day-Book of C.S. Rafinesque 1832 to 1834,*[6] The worth of twenty-two items of his material assets were listed—herbarium of 36,000 specimens, woodcuts, Pulmel and other medicines, manuscripts, maps, shells, fossils and minerals—$9,262. He also listed "old claims and bad claims" including money due to him from Transylvania University. Rafinesque's memory was long, for he also included old debts from France and even from Sicily, along with claims to his inheritance from his father and uncle that would never be collected. His cash reserves were always dangerously low—$260 on January 1, 1832, and $166 two years later. Nevertheless he was careful to pay his bills on time. Through 1832 and 1833 he earned $617.79 from the sale of Pulmel, his medical consultations, and his publications. He spent very little on food—only $98.15 in 1832, and $74.60 in 1833—numbers suggesting that he was an exceedingly frugal man, living on salt pork and cornbread, an ascetic dedicated to natural science and the intellect.

Rafinesque spent a small amount on wine but did not drink hard liquor, nor did he smoke. His condemnation of the smoking habit was impassioned. However, if a man must smoke he should smoke cigars made from a more fragrant leaf—Liquid-amber [sweet gum], sweet fern, and wintergreen.[7] Rafinesque never commented on the widespread American habit of chewing tobacco and spitting, which so appalled European visitors.

There is reason to believe, according to Pennell, that he was not always scrupulously honest in his dealing with his colleagues. Soon after the death of his good friend, Zaccheus Collins, in 1831, he submitted a bill for $340 to General Daniel Parker, Executor of the Collins estate. Rafinesque claimed that the bill was for specimens supplied to Collins and for publications extending from 1818 to 1831. The bill was actually $402, but Collins had advanced Rafinesque small payments over the years amounting to $62, which was subtracted from the $402. There were no meaningful records of the transactions (bills or receipts), and Collins was not there to verify or refute the putative transactions. Contested in court, three arbitrators rendered their judgment of Rafinesque's claim. The defense claimed that the specimens were not asked for, that no bills were ever submitted to Collins, that the small amount paid to Rafinesque was, in fact, a loan to him, and that "sales" of specimens were no more than unsolicited gifts to his friend over the years. Their argument seemed to convince Rafinesque's lawyer who suggested that he drop the matter, but Rafinesque persisted, and to everyone's amazement the arbitrators awarded him $173. However, in lieu of a cash award, Rafinesque accepted Collins' coveted, valuable herbarium.[8]

Another difficult situation arose involving William Wagner, a wealthy trader, enthusiastic promoter of the natural sciences, and founder of the Wagner Free Institute of Science of Philadelphia. Rafinesque and Wagner appeared to be friends, prompting Rafinesque to ask his colleague to be guarantor of a note for money that he desperately needed. This was refused outright, but despite the rebuff, in April 1840, just five months before Rafinesque's death, Rafinesque sent his good friend a letter requesting that he send $5 for his *Amenities of Nature*, or at least $1 for a pamphlet that had already been sent. Once Rafinesque made a friend, he presumed that as an impoverished scientist working for the good of humanity, he would provide the science and the specimens, and the friend would be obliged to help him in a material way. But Wagner was furious with Rafinesque's presumption, replying that he was returning the pamphlet unread and that he wanted no further literature from Rafinesque. In Wagner's final letter he wrote: "You again speak of giving for nothing your labor, your discoveries, etc. I never received any of your labors, or discoveries, or anything else. If you can show I have, send me a bill of it and I will pay your demands." Rafinesque's response was one of "sorrow" at the lack of appreciation. Cordial relations at an end, in his final letter to Wagner he became embarrassingly boastful about the famous European scientists who were about to purchase his specimens. Rafinesque's response to threat and insecurity was to attack them head on with bluster—transforming his most desperate hopes into reality. There is something reminiscent of the Collins affair here, but the difference was Wagner was still alive, able to confront Rafinesque and prevent him from reifying a confidence into a legal, financial contract. Still, Wagner seemed unnecessarily brutal with a hapless Rafinesque, reducing him to cinders—a merciless pummeling that would probably have been applauded by almost everyone of note.[9]

Whatever the rude treatment, Rafinesque was apparently undaunted, but it was noted that he seemed to be an unhappy man who never laughed. To make and maintain contacts, he attended meetings of the ANSP where Philadelphia's scientific elite, together with interested physicians, clergy, and wealthy businessmen conferred. Rafinesque was a large presence at these gatherings, for he was not shy about commenting after every speaker and expressing learned opinions on every subject. At a meeting in October 1825, Rafinesque impressed one participant as a most striking character. "He had a fine black eye, rather bald, and black hair, and withal is rather corpulent." He was reported as "eccentric and egotistical to the last degree," and it was said that, "he attempted to cover the whole field of science, history, and finance." His scientific works were for the most part ignored by his contemporaries, and in return he handled them without mercy.[10]

In 1832 he launched the *Atlantic Journal*, a periodical devoted to the historical and natural sciences, which he hoped would counter the conservative (Linnaean) *Journal of the Academy of Natural Sciences of Philadelphia* and *Silliman's Journal,* both of which excluded him. The *Atlantic Journal* was poorly received and lasted only two years. Rafinesque explained that his publication failed because it was "too learned and too liberal" in a land where there are "a crowd of literary journals . . . which contain hardly anything beyond plagiarism and vapid trash . . . that often succeed much better. I ought to have copied them to insure success; but I would not thus degrade myself. All my articles are written on purpose." He had written 160 original articles for his failed *Atlantic Journal.*

Attacks on Rafinesque's faulty scientific practice often bore a large measure of *ad hominum* nastiness. The most virulent attack ever made on Rafinesque—his integrity, judgment, and sanity—can be found in a paper written by G. W. Featherstonhaugh and published in his *Monthly Journal of Geology and Natural Science.*[11] Featherstonhaugh, a difficult, prickly, English geologist working in the United States, was a person with a remarkable number of enemies. Reviewing two of Rafinesque's works, he prefaced his diatribe by announcing that he had taken on the task of exposing Rafinesque as a service to natural science to protect the naïve and semi-ignorant, and to preserve the reputation of American science abroad, weak as it was. The works of Rafinesque he was reviewing were sufficient to condemn him.[12] In describing Rafinesque's writing he used such colorful terms as "off the perpendicular . . . the worthlessness of such a farrago as he has now let loose upon us," and then he went on to discuss specific details. Rafinesque was ridiculed for creating new taxa on the basis of a fossilized horn, jawbone, or tooth—a practice introduced and justified by Cuvier himself.

As for Rafinesque's *Atlantic Journal,* "we despair of doing justice to its various merits; it is a perfect museum of curiosities, and those who desire cheap amusement—for it only costs twenty-five cents—cannot do better than purchase it." He ridiculed the advertisements in the journal, and Rafinesque's profession as a "pulmist." He declared the journal's zoology, geology, history, and linguistics worthless. The material coming from his other works such as "that insane mass called *Annals of Kentucky,* and others of the same quality" were not new. Featherstonhaugh's diatribe was shocking. Words were twisted, venomous arguments that were logical but absurd, abounded—all used against Rafinesque in such a relentlessly sardonic way that their mere reading was an unpleasant experience, and one cannot help but feel sorry for poor, benighted Rafinesque, subjected to this public humiliation.

Although Rafinesque responded that it was beneath "the dignity of Science to imitate the example thus given us," he did become rather vitriolic in his outrage defending himself in the journal the *Atlantic Journal and Friend of Knowledge*, of which he was editor and proprietor. Aflame, he wrote of Featherstonaugh's attack: "From beginning to end a jumble of scurrility and a public attempt to injure us. This article is a disgrace to the writer, and the journal where it is found . . . nothing half so spiteful and disgraceful was ever before Stereotyped here or any where else . . . an absurd review." Rafinesque used such full-blown language as "a tissue of absurdities and false statements as this shameful rhapsody contains. . . . All his stones and bones are mere Feathers." Attacking Featherstonhaugh *ad hominum,* his integrity and his competence, he said: "If Mr. F. has been successful as a lecturer [Rafinesque claimed that his lectures were not worth attending], he has failed as an editor, a man of general science, and even as a Geologist. He has disgusted many persons by his proud and overbearing sufficiency. . . . As he is neither a Zoologist, nor a Botanist, nor a Philologist, nor an Antiquarian, although too proud to acknowledge it, he cannot understand my labors, and rails at them as ignorant men so often do . . . he has made no discoveries! while I count mine by thousands." Featherstonhaugh was a man without qualifications as an investigator in the field—the ultimate damnation according to Rafinesque.

George Harlan of the Academy of Natural Sciences of Philadelphia had introduced the two protagonists several years earlier, and both Harlan and Featherstonhaugh had had cordial dealings with Rafinesque. A difficult, arrogant man, Harlan had come to detest Rafinesque, and was most probably the instigator of Featherstonhaugh's scathing denunciation. Previously, Featherstonhaugh's had reviewed Rafinesque's work on fish and on bivalve molluscs in a most favorable way and had echoed Rafinesque's claim that others had unfairly assigned names to species that he had already discovered and named. Friendly reviews had left Rafinesque astonished and unprepared for Featherstonhaugh's assault.

The offensive was reinforced by an open letter from Harlan describing Rafinesque's lack of integrity in his dealings with his colleagues. To Rafinesque, Harlan was Featherstonhaugh's "sleeping partner. . . . The public shall easily discriminate between the plain truth, and their farrago of envy and spite." The convincing point-by-point defense, especially of the Harlan letter, was followed by a paper in the *Atlantic Journal* entitled "On the false Rhinoceroides of Featherstonhaugh and Harlan." Rafinesque damned them both as "credulous" and "ignorant" for making a "shameful blunder," because if they had had an ounce of sense, they would have realized that the fossilized jawbone of a rhinoceros they described was merely indurated sand, or stone.[13] They had made a diagnosis and had considered it a "great geological discovery," on the basis of a single specimen of bone, the same act for which Featherstonhaugh had ridiculed Rafinesque. It is possible that Harlan had turned on Rafinesque, because he had revealed Harlan's embarrassing blunder. The burden of having Harlan, who was so influential, as an enemy, was a serious matter—a virtual excommunication from the Philadelphia scientific community.

The damage done in these salvoes may not have been great, only because both Featherstonhaugh's and Rafinesque's journals were obscure and read by so few, but more important were Harlan's words in the introduction to his *Fauna Americana*, an influential work read by all biologists: "Mr. Rafinesque, Professor in the Transylvania University in the state of Kentucky, has described, or rather indicated, a great variety of animals, but his insulated situation, and almost utter ignorance of the labors of other naturalists, have seduced him into grievous errors, and occasioned much confusion in natural history. It is possible that some of his animals may be new species, but from the looseness of his descriptions, we have been obliged to reject them in almost every instance."[14]

One of Rafinesque's goals was to reach a wide audience of intelligent people and to enlighten them by integrating science into the everyday life of the citizen. Between 1827 and 1832 he published in Philadelphia journals (*The Saturday Evening Post* and *The Casket or Flowers of Literature, Wit and Sentiment*), under the rubric *The School of Flora,* ninety-six short illustrated articles, each describing a plant. The series was written in the spirit of the European "sentimental botany" of the time, and its intent was not only to inform and popularize but also to reveal to the reader the magic of flowers, their esthetic and poetic souls, and their "moral" nature.[15] In keeping with the Romantic spirit, Rafinesque was especially interested in "teaching the ladies." Beverly Seaton points out, however, that his attitude was rather negative toward women, sometimes stressing their inconstancy and bad temper, perhaps a reflection of his own experience with his former wife left behind in Sicily. In fact, as a boy he must have idealized women when he wept as he read St. Bernardin's popular *Paul and Virginie*, and as an adult, his criticism of them was in sorrow for their fall from grace. He seemed to be of two minds about women, often solicitous, addressing his remarks to them and imploring them to make a special effort to understand what he had to say, for it would be to their special benefit.

Rafinesque's later years in Philadelphia were spent writing, and were increasingly taken up with philosophic matters—sedentary activities—traveling less due to a lack of money, and nursing the few pennies he had. He organized his rich collections and tended to his correspondence with Cuvier, De Candolle, Brogniart, and Swainson, providing them with biological and mineral specimens. He also worked on his *Autikon Botanikon*, a list of new and rare plants, and he labored to further his "great manuscript work of Illustrations to all my travels and researches, containing 3,000 figures and maps in 30 volumes or Portfolios." This latter work, impossibly ambitious, never saw completion or publication. Over the years he was occupied writing such grand works as a history of the American nations (in "10 or 12 volumes"), selling rare books, biological and mineral specimens, and attending to his Divitial Bank which was finally established in 1835, a decade after it had been proposed. In the end, his banking enterprise must have been dealt a heavy blow in the serious Financial Panic of the late 1830s, and it certainly did not survive Rafinesque's death in 1840.

Confined to the indoors during the winter months, he converted his field observations into articles for publication in American or foreign journals. One such paper, based on his archaeological explorations of 1815–1833, described the remains of two sets of Indian villages in western Kentucky, one dating two thousand years and the other five hundred years of age—and he claimed that there were hundreds of such remains of habitations. Perhaps Europeans with a dim, romantic understanding of the New World might be impressed by Rafinesque as an intrepid explorer of picturesque but dangerous wilds—but who else would listen?[16]

When spring arrived, Rafinesque was drawn to the countryside, sometimes fleeing to escape exposure to cholera, which recurred with the arrival of warm weather. On several occasions he explored the valley of the Schuylkill River that ran through Philadelphia, following it up to the coal mines farther north—promising country for the geologist and paleontologist. Not only did he collect rare plants and minerals on these trips, but he also studied geological formations, hunted for fossils, and kept his eye open for artifacts of the Native American, all of which he shipped back to Philadelphia, to add to his extensive collections. In other years he explored the Pine Barrens and marl pits of New Jersey, so rich in fossils (including the remains of dinosaurs that were uncovered in 1858), western Pennsylvania, Delaware, New England, Maryland, and as far south as Virginia. Another expedition took him over the "mountains" of New York to Troy and the sources of the Delaware and Susquehanna Rivers, paying special attention to the physical geography of the region. In his travels he used any and every means of transport but mostly he walked. In 1832 he took his first railroad journey, from Philadelphia to Baltimore.[17]

Mostly he walked, a wonderful time of calm and solitude and an inspiration for his poetry as he scanned the earth's surface for plants, rocks, and fossils. Seeing so much, he must have developed a sacred sense of possession of all he beheld, for there were no others to share his immediate experience.[18] Rafinesque estimated that in all, he had covered six thousand miles on foot, sometimes over mountains. Since he was a remarkable conversationalist and a well-known authority in many fields, this spirited eccentric brought a little excitement to quiet, isolated habitations where he dined and lodged along the way. He brought information from afar and gathered the latest intelligence of the district. Other than his living off the land, his travels were never subsidized, and by his own extravagant estimate probably cost him as much as ten thousand dollars.

Although he merely touched upon the people he knew or visited in his *Life of Travels*, the list was extensive. He made a clear distinction between those he considered his friends (an impressive list), those he considered his enemies (equally impressive), and others who simply neglected his work (Lea, Say, Nuttall, Lesueur, Bigelow, Godman). In some instances his evaluations were faulty, revealing a willful blindness and a shocking inability to judge how people felt about him; some "friends" were really quite hostile. For instance, he considered Peter Steven Duponceau, Philadelphia's authority on Native American languages, to be a friend, but it was known that Duponceau considered Rafinesque's philological work nonsense.

After recounting his achievements until 1833 in *Life of Travels* and feeling that he had not met with the success due to him, he wrote plaintively: "Whether I will ever be able to turn them to some profitable account, either for myself or the public, is a matter of uncertainty. It is time however that I should spread before the public of both hemispheres, the results of so many years of labors and exertions. What I have already published are mere fragments, compared to what I may do yet, if allowed to or able to give the fruits of my historical researches and discoveries in natural sciences." A moment of uncertainty was quickly dispelled by unquenchable optimism, no matter how unrealistic.

As Rafinesque reached his fifties, he began to think about the family he had not seen in fifteen years. He had departed from Sicily in 1815 and had left behind his four-year-old daughter Emilia and a wife of questionable devotion. When his ship foundered off Long Island, his wife, thinking (or hoping) he was dead, married with unseemly haste Giovanni Pizzalour, a "comedian." Rafinesque tried unsuccessfully to have his daughter sent to him in 1816 and 1817, and then he lost all contact with his Sicilian family. Only after thirteen years did he bestir himself to make inquiries about their welfare—when his daughter would be a young woman of nineteen years. Rafinesque's quest to renew contact began with an appeal in 1830 to his sister Georgette and her husband Paul Lanthois in Bordeaux to find Emilia somewhere in Italy, but it was Rafinesque himself who found her through the offices of the French Consul in Italy. Emilia wrote a heartrending letter to her long-lost father: "My dear and loved Father I can not possibly describe to you the inexpressible pleasure which I have experienced in seeing for the first time your venerated handwriting, how many tears I have shed reading and rereading the tender lines dictated by your paternal solicitude. My dearest prayers have been answered. . . . Good-bye my dear father; your Emilia repeats every day your name. She will try with her conduct, with her tenderness, to cancel the past, to beautify the ending of her career. I beg you to send your paternal benediction, and with tears I kiss humbly your hands and call myself your tender daughter." All of her subsequent letters were written in this imploring, melodramatic vein.

Emilia had a long and sad story to tell. She, her mother, whom Rafinesque detested, and her mother's second husband had moved to Naples, where Emilia became an actress, and then to Rome, where she was courted by an Englishman, Sir Henry Winston, who deserted her after learning of her pregnancy. Emilia gave birth to a daughter Enrichetta. Life in the provincial theater was difficult and disruptive for the family, and it paid little. Emilia was the sole supporter of her family—her daughter, her mother Josephine, her aged stepfather, and her two young siblings. Emilia, accompanied by her mother, spent time touring in Malta and Tunis.

Communication between Rafinesque and his daughter continued, but his role in the relationship was ambiguous. Although there was something of the rich American patriarch in his claims, he was barely solvent (especially between 1832 and 1834) and unable to solve the problems of his destitute family. Despite his inability to help, he made the stern demand from afar that Emilia must decide between himself and her mother, whom she loved. She replied: "You impose upon me to choose between you and my mother, between being a lady or an actress. . . . How would I ever be able to leave my mother in poverty and live myself among comforts." Her assumption that her father was a prosperous, learned scientist was not discouraged by Rafinesque, incapable of admitting to his family his poverty and all its indignities. She was tortured by the fact that her father's abhorrence of her mother was unrelenting and that his opinion seemed misguided. According to Emilia, Josephine had been a loving mother who had brought her up as best she could, but Rafinesque demanded that only if she, Emilia, would abandon her mother and her family, would he make all arrangements for the voyage to America, where she and her daughter would live a life of ease. Emilia pleaded that some financial aid be given to her mother, stepfather, and two half-siblings.

Rafinesque was playing a game—a stern, judgmental father contemptuous of the theatrical world, trying to remove his daughter from a life of sin, to set her on the right path. But he was so far away, and he really could not afford to provide for his daughter and granddaughter at all. However, in his remarkable entrepreneurial fashion, he arranged for her to sing with an Italian Opera company in New York willing (if true) to pay all travel expenses for her and her daughter from Leghorn to New York, and pay her $150 per month. For reasons unknown, Emilia never came to America, and little more is heard of the matter. Rafinesque suspected that she was prevented from accepting this attractive offer by her family ("rapacious relatives"), but it is most probable that Emilia chose not to renounce her mother. In his will he divided his estate equally between Emilia and his brother Anthony Augustus's two children Jules and Laura, and wildly overestimating the net worth of his estate, he requested that some of the money be used to found an orphan school for girls.[19]

From 1830 until her death in 1835, Rafinesque corresponded with his sister Georgette for whom his *Life of Travels* was written. To her, filled with pride that her brother had won a Gold Medal from the Société de Geographie de Paris, Rafinesque was a hero in an America that was the land of fulfilled promises, and she advised him not to return to Europe where life was so difficult. She gave him news of Europe and he sent her samples of Pulmel and of his books, journals, and pamphlets.

By 1836 Rafinesque's myriad activities—his banking enterprise, medical practice, and especially the sale of Pulmel—had provided him with enough income to publish at his own expense much of the writing that had been accumulating in his drawer. Hitherto, his writing had earned him almost nothing, and if no one would publish his work, then he would publish it himself. In the following four years a substantial Rafinesque opus was printed—with as many pages as he had published since the beginning of his career. Several major works appeared in quick succession, beginning with *The American Nations* (in two parts totaling 552 pages), two editions of *The World, or Instability* (248 pages), *A Life of Travels*, *New Flora and Botany of North America*, and four parts of *Flora Telluriana*. The next year, further parts of *New Flora* and of *Flora Telluriana, The Universe and the Stars, Safe Banking Including the Principles of Wealth* (in all, 412 pages) were published. In 1838 Rafinesque printed another two parts of his *New Flora,* and additions to *Flora Telluriana, Genius and Spirit of the Hebrew Bible, Alsographia Americana, Celestial Wonders and Philosophy, Sylva Telluriana,* and *Ancient Monuments of North and South America,* a total of 1,032 pages, followed in 1839 by the *American Manual of the Mulberry Trees.* In his final year, Rafinesque, though ailing, managed to have printed at his own expense *Autikon Botanikon, The Good Book and Amenities of Nature,* and *The Pleasures and Duties of Wealth* as well as many smaller papers that were contributions to edited works, journal articles, and pamphlets—an unmatched wealth of diverse interests. Most of these publications are scarce today because they were printed in small numbers (usually much less than one hundred), on poor paper that rapidly deteriorated. Many of Rafinesque's published and unpublished writings were consigned to the rubbish heap after his death, but fortunately Samuel S. Haldeman, despite his having written a damning critique of Rafinesque's work, quietly salvaged several of Rafinesque's manuscripts. Material from Rafinesque's Kentucky days was also collected by Lyman C. Draper for the Wisconsin Historical Society.[20]

The World, or Instability of 1836 was a vastly ambitious poetic work of about six thousand lines, a "mirror of the world" divided into twenty cantos, each dealing with a cosmic, all-encompassing subject—the universe, the earth and moon, life and motion, love and sympathy, sublimity and the deity, religion, free will, angels and devils, mankind and society, peace and war, toleration and selfishness, passion and pleasures, wisdom and knowledge, arts and sciences, women and children. In the "Exordium" (Introduction), Rafinesque sang of the "laws of change and symetry[sic]," in which his views on biological change and descent of biological forms were expounded. The preface (or "notes"), supposedly written by editor Constantine Jobson, who introduced the poem to the public, could have been written by none other than the shameless Rafinesque himself, for it praised the work to the heavens. His epic poem, the "poesy of the Soul," was described as "an unusual literary effort. . . . This curious and moral poem is novel and unique: it bears the stamp of genius, which alone can strike a new path in poesy as well as philosophy." The poem was conceived and written in the style of Milton, Pope, and Erasmus Darwin and comparable to them. In some ways it was superior to their classics ("in moral tendency, variety of subjects, and sublimity"), and modestly, in other ways it was inferior—in language and polish—but it did contain "sublime" passages. He was in fact critical of Milton's poetry, but rendered flattering judgments on various sections of his own work: "The hymn to the firstborn of God is truly beautiful . . . the hymn to Peace, is but a trifle. The universal hymn and prayer at the end is excellent, and suited to all Religions. . . . Beauties abound in this poem, they are scattered like gems from beginning to end." The philosophy of the poem is "celestial and ethereal. . . . The little poem on women is delightful, and cannot fail to please the sex it extols; it ends by a happy transition to playful children." The "editor" conceded that works fashioned by the mind of a mere mortal must be imperfect: "Whenever great beauties are found in a work of Genius, corresponding defects may be expected," but these seem to be almost trivial and debatable, "forgivable imperfections" more in the mind of the critic than in reality.

Rafinesque's intent was "to prove that *Instability* is as much a law of nature as attraction or gravitation, that it rules both the physical and moral worlds, and that "even the Heavens are not stable." Religion, also "like everything else on Earth is subject to perpetual mutation and fluctuations," a worldview that derived from the Stoics through Seneca ("everything is forever changing"), and more directly from Lamarck. There was an underlying determinism in his thought, for he spoke of attaining perfection "in afterlives" after inevitable change and death. Truth was elusive, knowledge uncertain, and the wresting of order from randomness, difficult. To Rafinesque the word *mutation* was not used in the modern, genetic sense. It only meant change, and to him it seemed to signify deterioration, a speculation that touched upon the notion of entropy. However, mutations "happen so gradually as not to be very striking till after a long while," just as, decades after Rafinesque's death, Darwinians explained why the evolutionary process was indiscernible to humans, given their perception of time. It would appear that Rafinesque's prescient speculation about evolution and the occurrence of biological variety derived from the general philosophic principle of instability and change that he professed. However, Rafinesque's thoughts about biological evolution were hardly more than a flash of brilliant insight without elaboration or development.

Rafinesque's great philosophic trinity was *instability, perpetuity, and diversity,* really the description of an open-ended, steady state system, which shaped the foundation of his general survey of "the universe, the earth and mankind." What was expressed in this poem, he believed, was a new, universal law analogous to Newton's laws of gravity but presented in poetic form in the style of Lucretius: "It is as if Newton had explained his laws of attraction and repulsion in a poem, instead of a mathematical work." He perceived art and science, poetry and mathematics as alternative means of expressing and understanding the truths of nature.

The religious spirit embodied in the poem was generous and tolerant in that it "admits of all others." His was a religion of love and hope in which "all are in the right, who seek a God, and do not persecute," and that it "wars only against evil, strife and the human Devils." Rafinesque's Natural Theology, essentially secular, rooted in the Enlightenment, appealed to reason above all. It was intensely idealistic and high-minded, and far more encompassing than the Christianity of the various denominations of his day. He frequently spoke of a mighty God but Christ was never invoked, and conventional Christian dogma had little play. Rafinesque was expressing the commonly held views of many freethinking people of his time, and however ambitious and embellished, the tract was not profoundly original. His poetry, while mundane and awkward, did have merit—in its sentiment, visual imagery, and drive. What made his effort unusual was the astonishing breadth of knowledge that he brought to bear on his philosophy.

Typical of Rafinesque, he ended his introduction with a truculent, preemptive strike against his potential critics: "Those who may dislike this poem must have a bad heart, be exclusive in opinions, or fond of strife and discord. To them it is not addressed, since it deprecates what they hold dear. But the wise and good, the sensible souls, the friends of peace and mankind, and above all gentle women, must approve of it."

Rafinesque's ponderings, his grand attempts to set in order all knowledge for the enlightenment of mankind were colored by the times—one of promise and optimism, in the aftermath of the American and French Revolutions. His earlier speculations were of a strictly secular, utilitarian nature, but as he aged his thoughts were increasingly expressed in more conventional religious terms—a sense of reconciliation was discernable. When Rafinesque set down his beliefs, the usual Christian certainties were being weakened by a general loss of interest by the public, by the success and excitement of a burgeoning science, and by contemporary philosophic analysis of religion. Increasingly, scripture was being read as a collection of interesting stories, and the status of Jesus was reduced from divinity to a teacher of moral behavior. In this attenuated state of rigorous orthodox Christianity, a man like Rafinesque, a humanist, was prompted to conjure his own version of the meaning of existence, and he was especially eager to eliminate the cruelties of the Judeo-Christian doctrine in his system, perhaps because he was hurt by the society that was so identified with this orthodoxy. In the end, Rafinesque visualized a world of solace in which he could be comforted, to help him through his time of misery and pain.

The Genius and Spirit of the Hebrew Bible published in 1838, revealed Rafinesque's astonishing knowledge of the Hebrew language, going so far as to write his own Hebrew-English dictionary that he claimed did not exist. Recruiting etymology and close examination of the text, he attempted to discern the original, "forgotten" meaning of words, ideas, and truths, for only with this information could one understand "the angelic religion of yore." The "pristine sublimity and accuracy of the original Hebrew language" would reveal a religion that "agrees in everything with our modern Philosophy, Astronomy and Geology, as I will be able to prove." According to Rafinesque, Christians were sadly misinformed because "errors and mistakes of inaccurate or fake translations made long ago" were still being disseminated by teachers of religion whom he berated for not revising and correcting imperfect versions of the Old Testament, handed down for centuries. Once a faithful translation was in the hands of these teachers, they would no longer be "deceived and deceiving." For the same reasons he had unkind words for the children of Abraham: "Thalmudists or Rabinists are the modern Jews that uphold all the errors and absurdities of the Talmud with the Mishna and Targum, traditions and commentaries. . . . Therefore the pretentions of these modern Jews to have held the pure doctrine of Celestial Religion ever since Adam, Noah, Abraham and Moses, is not only absurd; but *totally false*, and those Xristians who travel on the same path are blinded by the same concert." Further, "as a nation [the Hebrews] had at least 10 great relapses into the Lap of Evil after the Adamic Fall."

Rafinesque's discussions, often rambling, not only encompassed biblical, secular, and divine philosophy, they were also concerned with the history of biblical times, Hebrew grammar, and the origin and variations of large numbers of words and names. As background, the basic precepts of all the major religions were presented, although surprisingly he remained silent about those of the Native American. Despite the array of scholarship marshaled to battle the sad history of fallacy, he failed to show how one goes from the old, established "misconceptions," to his revised, radically enlightened version of biblical precepts and his Celestial religion (discussed below). Apparently he felt he did not have to argue his case point by point. The mere presentation of his information would fully confirm his assertions, but in fact the connection between his revisions and the revelation of a Celestial religion was obscure,

Rafinesque having assumed that sympathetic readers would make their own connections and provide their own insights.

Celestial Wonders and Philosophy (1838) was a companion to *The Genius and Spirit of the Hebrew Bible.* Both were small books printed on such poor quality paper that the few extant copies are crumbling. Both were dedicated to, and printed for, the Central University of Illinois, an institution that was going to be established but for the moment existed only in Rafinesque's head. It was to consist of five colleges (Agriculture, Arts, Sciences, Medicine, and Medical Sciences) to form a university that was to open in 1850. *Celestial Wonders* was sponsored by the Eleutherium of Knowledge, a kind of open university founded by Charles Wetherill to edify anyone who desired enlightenment. According to Rafinesque, it was "a free and gratuitous college of Knowledge and Philosophy." Wetherill, a member of an old and prosperous family of manufacturers in Philadelphia, fancied himself a philosopher, and as a friend, and Rafinesque's patron, he probably supported his banking scheme during a period of hard times. The two men were copublishers of a book inspired by the writing of an eighteenth-century English antiquarian and historian, Thomas Wright, the work dedicated to "worthy minds and benevolent men."[21]

Celestial Wonders and Philosophy, written by Rafinesque in a week's time, attempted a synthesis of science, religion, and philosophy from which would arise an overarching "Celestial Religion." Its purpose was to unite "all Sects, Denominations, and Churches in a Single Universal Evangelical Religion of Love and Peace" over which God, who manifests infinite love, and his Angels, preside. Rafinesque believed he had attained the high ground, above the petty quarrels of organized religion and the various sects and denominations that he believed arose only because of ignorance and misunderstanding of language, as outlined in *The Genius and Spirit of the Hebrew Bible.*

Rafinesque stated that he was inspired by the scientific discoveries of many wise people, although Newton was hardly mentioned. Physics, astronomy, biology, and language were discussed in detail, because only with knowledge of the physical world—an indispensable element of Celestial Enlightenment—could one understand, and arrive at a "blessed state." He claimed, "[E]very pure Botanist is a good man, a happy man, and a religious man. He lives with God in his wide temple not made by hands." Dwelling upon problems of cosmic dimensions, Rafinesque felt that his imagination was set free to reflect on the musings of great minds and mystics of former ages.

According to Rafinesque's cosmology, the human Mind, which was distinct from matter, was preeminent, and responsible for the creation of the universe, along with Design and Volition. Wherever there was Mind (the active rays of the soul), there was "Order and Concord." "Planets and Stars are directed by motive intelligent powers . . . and there is Mind in them as well as Light and Matter." Despite his professing rigorous adherence to fact and logic, he erected a shaky and fanciful metaphysical edifice, and we are left more than a little bewildered by his turbulent, sometimes obscure flow of thought. His thinking was not always logical or linear, but sometimes his words were crystal clear: "The art of seeing well, or of noticing and distinguishing with accuracy the objects which we perceive, is a high faculty of the mind, unfolded in few individuals, and despised by those who can neither acquire it nor appreciate its results." He, of course, was one of the chosen few. There were two Rafinesques—the precise observer and classifier and the addled victim of his own imagination.

Rafinesque's celestial meditations on the many forms of religion bore a close resemblance to his thoughts about the varieties of plants and animals in nature. In religion, incessant change throughout time had resulted in the formation of numerous forms of worship; in nature, constant "mutation" had generated innumerable species and genera of plants and animals; both were evolutionary progressions. Rafinesque had a very sharp eye for multiplication of forms in nature, a phenomenon he accepted with wonder, but without moral judgment. On the other hand, change in religious and spiritual sentiment and belief, was one of inevitable deterioration, a corruption due to ignorance and error that led to hatred and strife, so evident in the world around him. In essence he attempted to create a taxonomy of religion, comparing its various forms (new species) as they mutated (and deteriorated) in time. He prescribed a return to a religion of Love and Harmony through faith in knowledge, reason, and the rational mind. Rafinesque's notion of a falling away from an ancient earthly paradise, a golden age of giants and gods, was common to many religions, both oriental and occidental, and there have always been prophets exhorting the masses to mend their ways and return to a pristine, blessed state. Rafinesque believed he was such a messenger, yearning to rescue the world from its despair. His pronouncements were hortatory in nature as he hurled a challenge to his various audiences informing them of their duties and responsibilities. He addressed "Wise men of the Earth! Worthy men of this age! Men of Wealth! And you Women!" For each he had a message, and then he ended his little book with a long invocation that sounded like a prayer from the Old Testament.

The *Good Book and Amenities of Nature,* an ambitious work printed for the Eleutherium of Knowledge, was really a return to his classifying efforts in the *Précis des Descouvertes* (1814) and *Analysis of Nature* (1815). Rafinesque attempted to categorize all knowledge, especially the sciences, and within this larger framework to provide a master classification for all plants and animals under the rubric *Somiology* ("knowledge of the organization of both plants and animals," a systematic botany and zoology), another expression of his name-devising talent. According to Rafinesque's system, all flora and fauna were each divided into ten classes, a classification based on morphology, which falls far short of modern, more broadly based taxonomic standards. Paclt (1960) has judged Rafinesque's scheme "both the first and the most ingenious system of descriptive biology."[22] An irrepressible compulsion to impose order on the world was once again manifest.

Over the years Rafinesque had published several accounts of the immense number of plants he had described and named, although he neglected to write a comprehensive account of the flora of Kentucky, a task for which he was particularly fitted. However, he did describe and name 612 Kentucky plants in various publications, and with great insight he divided Kentucky into four distinct plant regions. Very few of his names have held up, and so it may be just as well that a *Flora* of Kentucky was not undertaken. In 1836 his *New Flora and Botany of North America* appeared, which was "to supply the omissions and deficiencies of all the writers on our botany." Yet when a presumably definitive *Flora of North America* by John Torrey, Thomas Nuttall, and Asa Gray appeared in 1838, Rafinesque was almost completely ignored. It fell upon Rafinesque to review the book, which he found "a medley of good and bad things." Seemingly without guile, he was "happy" that Torrey had finally adopted the Natural System of classification after clinging to the outmoded Linnaean Sexual System for so many years. However "the natural order of this *Flora* are deficient in arrangement, precision, names, synonymy, and composition, their characters are vague, loose, incorrect, and unfit for study." Numerous errors and misnomers were listed, and plants well known to other botanists were omitted. As for Gray, Torrey's student who was to become a professor at Harvard University, he "is also quite a beginner, and taught by Torrey to judge of plants at a mere glance, without studying the characters." There were many oversights, and they had "forgotten many of my plants." After all the fault finding, Rafinesque ended on a mollifying note—"these remarks have not been written with any unkindly feeling." He considered Torrey, his friend, one of the best botanists in America, and he hoped that his criticism would help improve his *Flora.*

Rafinesque's *Flora Telluriana*, three parts published in 1837 and a fourth in 1838, is the magnum opus of his botanical work and his thought, embodying his dynamic notions about taxonomy that had really not changed since his *Analysis of Nature* of 1815. Discussion in the *Flora* had a strong historical bent and was most respectful of the great masters of the past, especially Linnaeus and Adanson, but he was merciless in his condemnation of several of his contemporaries—especially the most prominent such as Lindley and Hooker. Lindley, a prolific and widely read botanist, was treated with scorn—"what a confusion, and what a blunder!"[23] The intent of the book was generic reform, and to establish one thousand totally new genera. Since he postulated that new species and genera were constantly being formed, Gray snidely suggested that if this ambitious goal could not be reached at the present time because of a lack of existing genera, all one would have to do is wait a century for the aim to become a reality.[24]

Rafinesque wrote with much bitterness, as he tried "stemming the current of botanical errors and blunders." He apologized for the high price of his *Flora*. But he was forced to publish in small quantities (160 copies), because "few copies can be sold in America, where Botanists cannot duly appreciate it, and they must be sent to Europe, to be often exchanged instead of sold." Nevertheless he was confident that ultimately his views would prevail as the younger generation of botanists came to see the correctness of his views regarding the formation of species and genera; the definitions of class, genus, and species; and how and why plants should fit more naturally into his scheme of classification in preference to others. As commonly practiced, the classification of plants was in a "state of disarray," which his system could put to right. Wistfully, he expressed the hope that someday he would become the wise elder who would advise those who followed him to look at his work and revise as they see fit, and to "imitate my zeal, and be happy in the lovely state of flowers."

After a life of chronic insolvency, Rafinesque spent himself in this final disgorgement, and at the time of his death in 1840, in what legend holds was a garret, his total cash on hand was said to be about six dollars, with only a few dollars in the bank and his rent overdue. He was not only the victim of his own personal financial failures, he suffered as well during the last three years of his life because of the Panic of 1837 that led to the widespread failure of banks, throwing the nation into a profound economic depression.

Chapter 15

LAST DAYS

In his last years, Rafinesque settled into a vigorous routine, attending to his banking interests and writing and publishing accounts of thousands of specimens obtained from colleagues or brought from Kentucky. He roamed less frequently, but on occasion he could not resist the urge to flee into the wilderness, leaving his troubles behind. In 1835 he explored the Allegheny region, and the next year the Pine Barrens of New Jersey.[1] There is little record of travel after this, clearly a period of withdrawal into the city, struggling to stay afloat financially to feed his publishing hunger.

At the time of his removal from Lexington and arrival in Philadelphia, he was described as "a short man, stoutly formed, and very plainly dressed. He appears to be about 40 years old. . . . His head is somewhat baled [sic] and he combes his hair directly across, has dark eyes &c on the whole a pleasing countenance."[2] But calamitous years of physical and mental stress, constant battles, and poor nutrition had worn him down, and after a decade—three years before his death—he had become "a little 'muffy'-looking old man resembling an 'antiquated Frenchman' . . . an exceedingly gaunt and tremulous old man,"[3] and to others he appeared "weedy looking." Slowly, he withdrew from scientific society for there was no profit there, and one senses that for Rafinesque, the world had lost its spark.

Long after his death, after years of repudiation, there were some who devoted themselves to rehabilitating Rafinesque's reputation. Such a person was Richard E. Call, the first biographer of Rafinesque who began the assembly of a *Bibliographia Rafinesquiana*, and *Bibliotheca Rafinesquiana*—truly admirable works. His emotional account of Rafinesque's death was highly dramatic, a threnody that tugged at the heart:

> The closing scenes of this man are of the saddest nature imaginable. He lived

> in the most abject poverty on Race St. Philadelphia, in a garret, surrounded by his books, minerals, plants and other loved natural objects. He shunned the company of others and had no, or but few, real and tried friends. Scientific recluse that he was in these days, there were none to care for him and help him in time of want. His scientific wants were still strong, and he struggled along in the unequal battle with fortune in the face of a disease that had no relief save death. The end came in 1840, when alone in a crowded garret, in a poor quarter of the great city, he died of cancer of the stomach.
>
> Language can not adequately portray the emotions that arise as these words are written. Here was a man who for years had loved and wooed that coy goddess whom we call Nature; a man who had the soul to appreciate both her richness and her profligacy; whose varied fortunes, both in letters and in means, seemed as the details of a romance; he had at last paid the penalty of being a part of that same Nature. He died without a word to cheer him, without a tear shed for him. Rafinesque! The name has gone to every land where science is cultivated. Rafinesque! The name has been bandied about in jest and contumely by those who should have hailed him as a brother. Rafinesque! Dead! He yet lives and will live as long as plants shall be studied and classified; as long as fishes shall unwittingly fall in the net of the searcher; as long as the waters of the West shall give life to mollusks; as long as changing stream of fleeting or moving star shall bear a message to men. Long may the name of him who studied them all and loved them all and understood them all be revered by those who regard the labors of the pioneer![4]

Rafinesque left behind a wake of confusion and uncertainty about his intellectual endeavors, his personal life, and even the details of his death. The usual account related for over a century was a melodramatic tale, operatic, in its details in which a starving, unappreciated scientist dies alone in his garret, a pauper, whose body was saved in the nick of time from being sold by a greedy landlord for dissection by insensitive medical students and haughty physicians, an ignominious end for our hero—but one that would pay for the overdue rent. Novels were actually written in nineteenth-century Philadelphia about the horrors committed by grave robbers, unsavory people who would indulge in such criminal trade, and the ensuing nightmare of a medical student, obliged to dissect a human cadaver in his training, who finds the subject his recently deceased sister or father, sometimes not quite dead![5] (One can only imagine the tearful reunion.) Rafinesque was spared the indignity of dissection but nonetheless was thrust into a pauper's grave, which was all but lost, a story reminiscent of Mozart's lamentable death and burial in a muddy grave whose location was soon forgotten. Many years later, a wealthy Pennsylvania lawyer and academic, a revisionist regarding Rafinesque's reputation, heard of this tragedy, and found and restored the grave. When news of this reached

Transylvania University, they requested that his sacred remains be transferred to a specially constructed tomb on their grounds, and there he rests, revered to this day.

The legend was based on the slightest of secondhand accounts of Rafinesque's death and burial and was written by an anonymous *H.H.*, for the *Philadelphia Ledger.* The brief article was unusual for the number of factual errors it contained—at least fifteen—rendering it an unreliable curiosity. According to *H.H.*, who had heard the story many years earlier from the undertaker, Rafinesque was born in France (with incorrect dates), and was an orphan who was shipwrecked in Nova Scotia. The wooden cross over his pauper's grave bore the initials *S C R*, and he was tended by his only friend, the late Dr. James Mease.[6] This distressing story, which had been endlessly repeated, was initiated and firmly established by Thomas Meehan, a publisher with botanical interests who was sympathetic to Rafinesque; he had known people who knew Rafinesque. In 1883, his imagination exercised, he wrote: "His chief home was in a dingy garret, with scarcely a loaf of bread to eat. He worked for science, as he understood it, to the last. He died on a cot with hardly a rag to cover him, and without a solitary friend to stand by him in his last hours."[7] The theme was taken up by Call, who with other earnest revisionists wanted to emphasize the cosmic injustice suffered by a misunderstood genius. While they emphasized the cruelty and unfairness of it all, they finished the tale on a hopeful, upbeat note, for now the truth was revealed—now the world understood!

A more realistic account was proffered by Pennell, who sifted through the holdings of the Academy of Natural Sciences of Philadelphia and came across documents given to the Academy by a descendant of S.S. Haldeman, the man who despite writing a disparaging obituary of Rafinesque, bought some Rafinesquian memorabilia and publications at a public auction after Rafinesque's death. In the Haldeman cache was a published medical history of Rafinesque's last illness.[8] Recent evidence uncovered by Charles Boewe also suggests that the standard story of Rafinesque's end was inaccurate in numerous important details.[9] Nevertheless, an aura of desolation during his last days, if not years, is palpable, and cannot be dispelled.

Rafinesque died in 1840 of a gastric carcinoma that had metastasized to the liver. However, his affliction was not a long, drawn out affair. According to a published report, he had suffered from constipation throughout the previous winter, and in mid-June he had bouts of nausea and vomiting with pain after eating. In late August he noticed a yellowing of his skin and a mass in the upper right quadrant of his abdomen, which probably caused the obstruction that was responsible for the nausea and vomiting. He rapidly weakened and died within a few weeks at 9:00 P.M. on September 18, 1840. Rather than dying alone and friendless in a garret on Race Street (between Third and Fourth Streets), Boewe states that he lived and died in a rented house at 172 Vine Street and that he was attended daily until the end by a very able Philadelphia physician, Dr. William Ashmead, who called in consultant Dr. Edward Hallowell. Both were graduates of the University of Pennsylvania, were members of the Academy of Natural Sciences of Philadelphia, and were interested in science and natural history. It is probable that they were especially attentive and concerned, because they knew about Rafinesque as a scientist and gave him the best possible care available, however ineffectual; he would have fared no better in a hospital. In accordance with modern medical practice recently introduced from France, Dr. Hallowell performed a very professional, detailed autopsy thirteen hours after Rafinesque's death, and he confirmed the clinical diagnosis. Rafinesque's entire history, pre- and postmortem, was written up and published by Dr. Hallowell in an age when medical people seemed to have no compunction about performing an autopsy on friends or even relatives because it was all in the interest of science.[10]

Characteristically, Rafinesque made judgments about his treatment. We learn that he refused to take calomel, a commonly prescribed but toxic mercury salt concoction that he had always wisely scorned as worse than useless. Being in considerable discomfort with bedsores, unable to hold down food, and probably in considerable pain, mercifully he was heavily sedated until the end. It would appear that Rafinesque was not entirely alone in the world, for he was probably tended daily by not only Dr. Ashmead but also someone who looked after his basic nursing needs. He was also visited by a druggist, by botanist Elias Durand, and by the well-known Dr. Mease, who attended Rafinesque as a friend, not as a physician, and later became executor of his estate.

Until a few months before his death, Rafinesque seemed to be as active as ever, planning future activities. His writing in the spring of 1840 has been described as "firm and steady," and letters written by Rafinesque a few months before his death to his friend Swainson was in a hand that showed "no sign of decline of his physical and mental vitality," and was much like his very first letter written decades before—requesting favors and information, and offering specimens.[11] In his final year he wrote the *Autikon Botanikon* and the *Good Book and Amenities of Nature,* serials in several parts, which suggests a plan for the future. Publication would also indicate that he was not completely destitute, because it entailed a considerable expense. He had made a modest amount of money in his savings bank venture, and he had been partially supported by Charles Wetherill.[12]

The received story is that after Rafinesque's unattended death, the landlord, and perhaps other creditors, refused to allow friends to give the deceased a decent burial, and so they locked the body in a room until it could be sold for dissection. An old friend of Rafinesque, Dr. James Mease, learning of his death contacted an undertaker, Mr. Bringhurst, and suspecting that the body was to be sold, forced open the door of the room (presumably where the autopsy had been carried out), and removed the body, lowering it with ropes from a window. In fact, the evidence is weak that Rafinesque's remains were to be sold for dissection (a degradation that would certainly have added to the pathos of Rafinesque's sad end in an age of Romantic excess). In accordance with Dr. Mease's plan, the body was buried in Ronaldson's cemetery, also called *Stranger's Ground,* established in 1827 for the respectful burial of visitors to the city like Rafinesque, of Protestant background but without church affiliation—a beneficent gesture.[13] The cemetery occupied the southwest corner of Ninth Street and Bainbridge Street, and according to Boewe's sleuthing, it was a "show place" with "plantings and gravel walks"—a park. A half-century later, however, the area had become a slum, the cemetery was unattended and overflowing with bodies, giving rise to the story that Rafinesque had been buried in a "potter's field."[14]

The burial, which was probably a small, lonely affair without the benefit of clergy, was paid for by the executor Dr. Mease with money that came from the sale of Rafinesque's worldly goods. A walnut coffin had cost seven dollars, winding sheet and shroud three dollars, hearse and carriage six dollars, all pointing to a rather minimal funeral. Pathetically, less than seven dollars was in Rafinesque's possession at the time of his death, and after the sale of Rafinesque's property for $131.42, and payment of expenses and commissions, Dr. Mease was out-of-pocket for $13.43. Rafinesque's treasures, his books, and unpublished manuscripts were considered junk. His books, which he had estimated were worth $1,250, were bought along with his clothing for $22.29, and even after this sale there remained eight cartloads of books and specimens. Also included were many of Rafinesque's unpublished manuscripts, which were sold for $5, some falling into the hands of S.S Haldeman, who later presented them to the Academy of Natural Sciences of Philadelphia—the remainder was rubbished.

Rafinesque's herbarium at one time supposedly contained some 50,000 specimens, but after his death it was found that part of the collection had been destroyed by rats and by neglect, and much of it despoiled by Elias Durand, a

Frenchman who spent many years in Philadelphia as a pharmacist and botanist. He had bought the entire collection at auction, mainly to obtain the specimens of Zaccheus Collins purchased by Rafinesque, and after destroying almost all of Rafinesque's material, returned to France with his herbarium, which he donated to the Musée d'Histoire Naturelle in Paris. John Torrey, a friend of Rafinesque, commented that Durand had been "agreeably disappointed to find in it quite a number of gems and duplicates, enough of many good plants to supply some of his friends," and as a result, the Durand Herbarium in Paris is only a shadow of the original Rafinesque herbarium, containing almost none of his numerous specimens of grass (a specialty of his). These specimens, in whatever state they were and however inadequate their descriptions, could have served to legitimize much of Rafinesque's work as his type specimens—perhaps thousands of them. They were lost forever.[15]

In 1833 Rafinesque wrote his will, a grandiose affair that reflected his utter inability to correctly evaluate his status—social, scientific, and financial.[16] Codicils were added in 1835 and were registered, with James Mease as sole executor. Mease does not seem to have been conscientious or effective, for the terms of the will were broken. Rafinesque's possessions were dispersed by public auction rather than by private sale as stipulated, and most of it was junked. Rafinesque believed he was a man of wealth, compelled to think carefully about how he would dispose of his worldly goods for the greatest benefit to mankind. In a kind of preamble he commended his "immortal soul to the Creator and Preserver of the Universe" and then departing from the usual, "the Supreme ruler of millions of worlds soaring through space, to be sent to whatever world He may deem fit and accordingly to His wise laws." He requested that his body be cremated, because he did not want it to "contaminate the earth and be the cause of disease to other men." However on his deathbed he changed his mind about this and other provisions of the will—and so he ended up occupying several square feet of the earth's surface in close company with strangers.

His personal property was left to his sister Georgette Louisa Lanthois of Bordeaux, and his daughter Emily Louisa, and with passion he justified his Draconian justice of cutting off his former wife Josephine Vaccaro without a penny. His books, maps, engravings, and collections of natural history were to be sold to pay for the printing of his manuscripts, maps, and sketches, with profits from the sales to be divided between his sister and daughter. His gold medal was to go to his nephew Jules (son of Antoine Auguste), to be kept with pride within the family.[17]

Estimating that his books, collections, manuscripts, patents, and inventions should fetch about $10,000, he instructed that a school for female orphans be established along the line of Stephen Girard's school for male orphans. Girard, a remarkable Philadelphia banker and merchant of French origin, was in his time the richest man in America.[18] Alas, the net worth of Rafinesque's estate turned out to be close to zero.

By 1914 a wooden cross marking Rafinesque's grave had rotted, and only after careful study of cemetery records by A.M. Hance and Samuel N. Rhoads, a bookseller, could the plot be located. Henry C. Mercer, a wealthy lawyer, philanthropist, archaeologist, museum builder, and tile manufacturer from Bucks County who had heard about Rafinesque, encouraged the detective work, and offered to place a stone monument over the grave.[19] However, the picture was complicated because each grave site was the repository of more than one body. At first it was thought that there were two bodies in Rafinesque's grave lying side by side, although it is difficult to imagine how this could be because graves were only four feet wide and Rafinesque was known to have been buried in a coffin. Further delving revealed that the common practice was to bury the dead serially, as deep as thirty feet down, and the records showed that there had been six burials in Rafinesque's grave site. Which remains were those of Rafinesque? Perhaps most people would be satisfied with the certain knowledge that only God knew which were Rafinesque's remains, and would let it go at that. However, befitting his genius for generating puzzlement, there arose a pressing need to establish precisely the identity of his remains.

Mercer was willing to disregard this complicating factor and design a memorial for the grave. His company, Moravian Pottery and Tile Works of Doylestown, Pennsylvania, created a slab of fine concrete that covered the entire site, bearing an inscription that echoed Rafinesque's words and sentiments:

> *Honor to Whom Honor is Overdue*
> CONSTANTINE S. RAFINESQUE
> Born Constantinople 1783
> Died, Philadelphia, September 18, 1840
> To do good to mankind has
> ever been an ungrateful task
> The works of God to study
> and explain
> Is happy toil and not to
> Live in vain
> This tablet placed here September, 1919

But Rafinesque's bones still could not rest in peace, for there were constant rumblings about razing and transfering the cemetery to a distant site so that the urban property could be used for other purposes, an appropriation that did not, in fact, occur for another thirty years when the land became a public playground.

In 1923, a Mr. Robert Spencer visited the site, and, learning of the possible destruction of the cemetery, informed his aunt Mrs. Charles F. Norton, a librarian at Transylvania University, about the fate of Rafinesque's grave. Mrs. Norton considered Rafinesque to be one of Transylvania's most famous faculty members,[20] and in fact the university was the only institution with which Rafinesque was ever fully affiliated. Mrs. Norton initiated a movement to bring Rafinesque's remains back to the university in Lexington where he belonged, where he would be revered by generations of students—a plan that was very well received by the university authorities. Gaining permission for Transylvania representatives to examine the grave and to remove Rafinesque's bones should have been a routine task, but it proved to be a complex tangle of administrative obstacles, a black comedy in a way that brings to mind the graveyard scene in *Hamlet.* In the course of the process, they learned that there were nine bodies in the grave. Perhaps Ronaldson's cemetery was truly a catchall for paupers.

Exhumation took place in 1924, the osseous fragments of the first body were reached six feet down and were set aside. The second set of remains were just below and were in a coffin with a collapsed cover and a bottom that was so solidly imbedded in the ground that digging stopped. It was decided that this was what was left of C. S. Rafinesque. However, Boewe, who described the entire affair in exquisite detail, has brought convincing evidence to bear that Rafinesque lay at least one level deeper and that the bones that now lie in the Transylvania vault are those of a woman, one Mary Passmore, who died of tuberculosis in 1848. The skull in the coffin was reported to be intact, but Rafinesque's skull had been opened during autopsy to examine the brain.

As Boewe points out, the transfer of human remains when ancient cemetery land is expropriated for another purpose can be a helter-skelter affair. If Rafinesque's bones were scooped up, they are now miles away from Ronaldson's, forever mingled with those of his fellow citizens. If they lay deeper, they still remain under the grass and asphalt of a playground, and so both Rafinesque and his father, Francois Georges lie in unmarked, Philadelphia graves.

The putative remains of Rafinesque were brought to Lexington in the spring of 1924 and were eventually placed in a raised tomb covered with the inscribed cement slab created by Mercer. The dark, windowless crypt in which it lies is located in the bowels of a Greek Revival building, "Old Morrison," on the campus of Transylvania University and is protected from the public by a solid door and two "creaking iron grills."[21]

David Starr Jordan, the major force in Rafinesque's rehabilitation, had written years before that Rafinesque had been so reviled, even the site of his earthly remains were unknown "after his long journey." He was now gratified that Rafinesque's final resting place was established and honored.[22]

The spirit of Rafinesque suffuses the vault. Known to be different and even mysterious, the presence of this Merlin, a man of occult powers and mystery, can be felt by anyone close to the source of emanations—certainly by students with a penchant for unusual pranks, for on one occasion Rafinesque's coffin was found on the chapel platform, and on another occasion during Parent's Weekend, a skeleton, purported to be Rafinesque's, was found seated in the chapel. Rafinesque has become a kind of patron saint of the institution, his remains treated as sacred relics, his tomb the highlight of a guided tour of the school where the curious may refresh themselves in a coffee shop called the Rafskeller. A day close to Halloween has been set aside at Transylvania as *Rafinesque Day,* in which two boys and two girls are privileged to spend a night of terror in the crypt, and the campus is lit by a bonfire around which students dressed as undertakers carry a coffin and scream obscenities.

The legend of *Rafinesque's Curse,* which has found fertile soil at Transylvania, began with Rafinesque's *Life of Travel,* in which he wrote that he had left the University in 1826 cursing President Holley and the school itself, and befitting someone of his powers his curses took effect, for Holley died within a year, and the school burned to the ground (not quite true). With mock ceremony, students and local newspapers have recognized Rafinesque's unique dominion.[23]

And so Rafinesque's tortuous, confused, and confusing journey continues, even after death. The earthly remains that devotees have entombed and revere are almost certainly not his, and yet the legendary powers he was believed to possess are still at work; he has entered into local folklore. Where is that energetic truant now? Rafinesque the man is no more, but his spirit, his indominability and insatiable desire to *know* and to *expound,* touches us all, and the sum total of human thought and knowledge is greater for his having spent a mere fifty-seven years on earth. At the same time, we can only shake our heads in wonderment at this strangely hapless, and utterly imperfect, extraordinary human being.

EPILOGUE

What are we to make of Constantine Samuel Rafinesque? Assuredly, he is unique in the annals of American Natural History, impossible to fully understand, especially if his writing has not been read in all its breadth. Considered judgments about him have vacillated over time, but few have ever doubted his brilliance and his capacity to digest immense amounts of information. With much truth, Ewan has written of him that he was "the most enigmatic and controversial figure in American Natural History."[1] The suggestion has been made that the word *rafinesque* should be introduced into the language, a companion to *picturesque* and *grotesque*—it might be useful.[2]

Rafinesque's energy, doggedness, and impatience to discover, name, and impose order on Nature enabled him to achieve prodigiously in so many fields—he can hardly be accused of being limited or repetitious. He lived at a time when science and technology were just beginning to show the promise of new worlds and new possibilities to come, and feeding on this ferment was this eccentric visionary, with a natural affinity for America—an unknown land of mystery, so rich in life, that was unknown to Europeans. Though not a political activist, he was fully aware of the rigidity and the abuse of power in Europe. He brought to the New World a view of life that would make him a liberal democrat in today's parlance—his concern with race, abhorrence of slavery, advocacy of the Native American, and his quest for social justice in medicine and banking.

We prefer heroes whom we can admire wholeheartedly, whose minor faults can be accommodated and may in fact impart a little color, complexity, and interest to a luminous image of a gifted spirit. With such subjects, there is always the lurking danger of turning biography into monotonous hagiography. But an extraordinary person with devastating imperfections confounds us and gives rise to turmoil in our critical evaluation of the good and the bad. Modern sensibility almost demands imperfection in biographical subjects, and in recent years there have been an abundance of such written lives. Are the heroes any less "heroic" for all their flaws?

A recurring pattern of performance is evidenced in almost every field in which Rafinesque delved, witnessed in virtually every chapter of this biography. Rafinesque would enter the scene eager to overwhelm with an impressive amount of his own data that unsettled those already working in a particular field. He would then publish a flood of bold assertions about the subject, which upon close examination were found to be overblown and faulty. Later, when the data itself was challenged and could not be verified, chaos ensued. Rafinesque would be criticized (or ignored), which incited his wrath, and grand polemics ensued. Whatever field he entered—triumphantly—he seems to have left in shambles, with established workers trying to put together some sort of order in their discipline, and he, himself, a victim of self-inflicted misery. But strangely, years later, Rafinesque, like the Phoenix, seems to have risen from the ashes, brought to life by a new set of admirers.

Anyone who attempts a comprehensive, critical analysis of Rafinesque must tread cautiously. Defenders of Rafinesque's name do so by favorably evaluating selected areas of his activities while turning a blind eye to the rest. Despite the fact that so much of Rafinesque's work is to be admired, there is an alarming amount of pure nonsense and—worse—fraudulent claims. His admirers seem to look uncritically at very limited facets of his productions that were perhaps born of too much forgiveness and sympathy for this dazzling man beset with so many troubles, while his critics relentlessly examine his work looking for errors, which are not difficult to find and do not stand up to critical analysis—and yet what remains is extraordinary.

Writing this man's life has been an exhilarating adventure. I have been carried along on a wild chase, at times in awe and at others exasperated, and still at others horrified by what Rafinesque did and by what others did to him. I found myself engaged in argument with him, amazed that such an intelligent man could behave so idiotically and could be so credulous as to defy elementary common sense. At times I found myself making excuses for him, losing patience but remaining sympathetic—I could not abandon him. I found Henricus Quatre's [Leon Croizat] unrestrained hatred of Rafinesque wrongheaded and deeply offensive,[3] but I must admit that as my view of Rafinesque shifted again and again, my inconstant, fickle judgment of him gave rise to feelings of private guilt and embarrassment.

Rafinesque belongs to a pantheon of remarkable fantasists whose membership includes Emmanuel Swedenborg and William Blake, men who flourished in the late eighteenth and early nineteenth centuries—Rafinesque's time. All were firmly rooted in this world—Swedenborg a successful engineer, Blake an engraver and typesetter, and Rafinesque a natural historian, banker, and medical consultant, but all are remembered for their wondrous creations of the imagination in religious philosophy, poetry, history, and science. The lives of Blake and Rafinesque have much in common. Both were deeply religious but shunned the traditional church, creating for themselves rather complex metaphysical systems relating man to the universe. The rebuff they suffered in this world gave rise to visions of a better world that was free of evil and pain, one in which they basked in the beauty and sanctity of Nature, and although life's experiences had battered them, they refused to succumb. Blake, a depressive, differed from Rafinesque, who despite having equal reason to be disheartened did not appear to have any seriously downhearted periods. Both had in common manic drive and astounding creative energy that sparked their explosive productivity. In the end, both died in obscurity, the identification of their remains either lost forever, or in doubt, attesting to the careless, unheralded burials they suffered.

Rafinesque's peers disliked him for both personal and professional reasons, and near the end of his life, mortally ill, this scruffy man was living in misery and almost friendless—mourners were hard to find. Desperately in need of institutional backing, he met with rejection. Young Asa Gray, the premier and most influential botanist of the next generation, wrote so disparagingly about Rafinesque in his obituary—his scholarly, somewhat sanctimonious diatribe codified all anti-Rafinesquian sentiment. He began with uncharacteristic carelessness—Rafinesque was "a Sicilian by birth," and later stating, "A gradual deterioration will be observed in Rafinesque's botanical writings from 1819 to about 1830." Pointing out numerous mistakes in his work, he mocked him and repeated the old, discredited joke about Rafinesque's mania for classification—extending to lightning! If he had read Rafinesque's short paper on the subject he would have learned that the accusation was false and that he was perpetuating a slander.[4] But Gray had a mission, and the subtext of the essay was that this unconventional foreigner was not a true representative of American science and that Europeans should not judge American efforts from his publications.[5] Gray's sentiment is repeated *ad nauseam* in numerous articles by other American botanists in the next half-century, for as the czar of American botany of his time, he brooked no challenge to his view. Both Gray and Samuel S. Haldeman of Philadelphia had written systematic, detailed criticisms of Rafinesque that were widely read, but more damning still were a host of private comments by colleagues in their letters to one another. Although they were not revealed to the general public at the time, they must have been truly representative of how Rafinesque was regarded by many of his peers.[6]

Still, one cannot but wonder why poor Rafinesque was so relentlessly reviled, for much of his work was no worse than that of some of his reputable contemporaries, and in fact, has well stood the test of time.[7] In the first half of the nineteenth century, the United States, recently independent, was a client of Europe, both economically and culturally, and its population was much smaller than that of any of the major European nations, but it was growing rapidly because of European investment. Notwithstanding national pride and jingoistic utterances, its scientific and cultural achievements and its institutions could not stand comparison with those abroad. American scientists, aware of their status, looked to England, France, and Germany for approval and as models to emulate. How could it be otherwise when almost all the books they read were by Europeans, while European experts rarely consulted or knew about American science and art? American students in England were known for their tiresome "touchiness" with regard to their nation's creative power, and the more established American scientists (Gray, Henry) instituted an informal policing mechanism to prevent less reputable and shoddy work by their fellow countrymen from ever reaching Europe lest it shake the fragile reputation of American science. The European opinion of the United States was already suffering from the instability, unreliability, and dishonesty of American business and banking practices—even Rafinesque was sensitive to this. The result was that although the publications of American scientists were often of good quality, they were at best cautious efforts that lacked inspiration and daring. Investigators eagerly made use of the flora, fauna, minerals, and fossils peculiar to America, obviously more available to Americans than to Europeans, but they preferred to publish stolid, imaginative papers of a descriptive nature rather than challenging works that deviated from the received opinion. Publications such as these could be subject to criticism, better dull than daring; correction of a European's error was the highest accomplishment.

According to their lights Rafinesque was the last kind of scientist America needed—an unreliable wild man—so American scientists attempted to silence him, discredit him, lest he come to be known as one of their authentic representatives. A cohesive network of established naturalists held sway over an impoverished, odd-looking Rafinesque, who spent most of his days in the field. The positive glee they felt upon hearing of Rafinesque's shipwreck and loss of all his possessions, the regret that he had not perished as well, followed by the virtual desecration of his worldly goods after his death, is shocking.

By contrast, the apotheosis of another foreigner, Louis Agassiz, helps us to understand the derogation of Rafinesque. Americans were proud of the fact that Agassiz, a young, articulate, socially adept Swiss scientist whose accomplishments were highly touted, actually *chose* America as his home, and in coming to the New World he elevated the level of American endeavor, just as European refugee scientists of the 1930s gave enormous impetus to American scientific creativity. Surprisingly, Agassiz was generous in his estimation of Rafinesque: "I am satisfied that Rafinesque was a better man than he appeared. His misfortune was his prurient desire for novelties, and his rashness in publishing them . . . Tracing his course as a naturalist during his residence in this country, it is plain that he alarmed those with whom he had intercourse by his innovations and that they preferred to lean upon the authority of the great naturalist of the age [Cuvier]—who knew little of the special history of the country—rather than to trust a somewhat hasty man who was living among them and who had collected a vast amount of information from all parts of the States upon a variety of subjects then entirely new to science." [8]

By the second half of the century, the United States continued as the land of promise but was beginning to show its industrial might, and it took great pride in the ingenuity of its radical inventions. Now equal to the population of European countries, after the Civil War the wealth of the nation had soared. Rafinesque's work was reevaluated by scientists who were born into a burgeoning, prosperous nation, confident in their sense of achievement and in their boundless faith in an American future. Indeed, Rafinesque's reputation began to peak during the Second World War when America truly came of age as the dominant world power. Americans could now afford to admit to his (and America's) imperfections, and it was quite acceptable to find that he really was a remarkable man. Rafinesque's story is one of savaging and salvaging.

Rehabilitation began in earnest a few generations after Rafinesque's death, when ichthyologist David Starr Jordan looked into Rafinesque's work and found much to admire. Jordan was followed by a small army of sympathetic admirers, biographers, and bibliographers, who by their impassioned, revisionist rhetoric swung opinion in the opposite direction. Instead of scurrilous depictions of Rafinesque because of his "occasional amours and inebriaties" who scandalized the pedants of the last century (utterly false), he was now an "erratic genius" who "lived a century too soon . . . the most gifted man who ever stood in our ranks . . . the greatest of them all . . . among the immortals of American Science." Even Asa Gray had to admit that Rafinesque was "a highly gifted man, far in advance of other writers in the botany of this country."

Why did most people come to admire Rafinesque decades after his death? The answer may lie in the fact that knowledge and thinking had advanced, and the burning questions of Rafinesque's day were either resolved or had disappeared. Scientists and natural historians of the late nineteenth century now had some balanced view of the problems of his day and could dispassionately judge whether Rafinesque's musings were correct, incorrect, or just ridiculous. They could properly evaluate the merit of his thinking without personal involvement, and without harm to their own reputations or that of the nation; the confusing and the incorrect did not pose a threat. On the other hand, his contemporaries, in the thick of their investigations, were uncertain and puzzled about the truths to be uncovered. The data had to be reliable, and there was reason to believe that Rafinesque's information was not solid, that his theorizing was not useful, so that they caused much confusion and turmoil.

Americans had a blind faith that science and invention, the keys to progress, would lead them into a better world, and Rafinesque himself ardently believed this. After discussing a curious veterinary problem and making suggestion about a cure for hypersalivation by horses, he declared: "It is my wish that these facts, conclusions, and hints may become useful, since the constant aim of science should be to apply its extensive resources to the practical benefit of our fellow beings. And such, I trust, will always be the ultimate objects and results of my pursuits." More generally, Rafinesque affirmed, "Every science is connected with the wants of mankind; and many sciences are indebted for their origin to those wants."[9] and firmly subscribed to the notion of progress and the perfectibility of humankind that was so much a part of the American ethos.[10]

In a prepsychiatric, pre-Freudian age, Rafinesque's bizarre "eccentricity" was dismissed as "the outcome of boundless enthusiasm for the study of nature," but more than one of his peers stated, not in a joking or colloquial sense, that he was insane. On the basis of reading Rafinesque's *Life of Travels* and a few other papers, a psychiatrist, J.M. Woodall, judged him to be sane, a compulsive, paranoid neurotic genius, and found no evidence of outright dishonesty.[11] Woodall probably had not read much of his writing nor his wild historical creations, and he could not have known about recent evidence that the *Walam Olum* was a fraudulent work.

At the end of the twentieth century, Rafinesque might have been diagnosed as suffering from a bipolar, predominantly manic disorder—chronic hypomania (mild mania), not violent, and therefore fully capable of functioning outside a mental institution, but becoming highly irritable and aggressive when challenged. Further, there were times when he seemed to manifest schizoid and paranoid tendencies.[12] There is little evidence that he ever experienced serious, depressed periods, attested to by the statement near the end of his *Travels* "I have often been discouraged but have never despaired long."

Rafinesque's complex behavior, puzzling to all, may not only be ascribed to a manic disorder but also to a condition known as *Narcissistic Personality Disorder*. The hallmarks of this syndrome are: "grandiose sense of self-importance; exaggerates achievements and talents; preoccupied with fantasies of unlimited success, power, brilliance, beauty or ideal love; believes that he is 'special' and 'unique' and can only be understood by, or should associate with other special or *high*-status people; requires excessive admiration; sense of entitlement (unreasonable expectation of special treatment by others); interpersonally exploitative, uses others to his advantage; lacks empathy; envious of others; arrogant behavior and attitude."[13] Without question, Rafinesque manifested most of these behavioral characteristics and this could account for his relentless self-promotion.

He could operate effectively with incredible energy and persistence within a rational, scientifically acceptable framework, and only occasionally did he reveal underlying psychopathology when he ignored or grossly violated the accepted values of society and the bounds of reason. He seemed particularly rational and reasonable when the welfare of people was involved. Although his behavior was "normal" much of the time, and indeed, he could be a kind and charming companion, there were periods under extreme stress when his actions and utterances betrayed full-blown mania, when he was excitable, paranoid, delusional, and even hallucinatory. Close examination of his work and behavior suggests that between episodes of demonstrable, overt abnormality, he operated at a steady, elevated level that would be considered merely eccentric, perhaps expansive and grandiose, rather than psychotic—a critical equilibrium sustained through chronic stress and frustration.

Whatever the mental derangement Rafinesque suffered, its expression was aggravated by his poverty and the rejection by his colleagues. However, his manic energy and singular focus were instruments of survival and the basis of his creativity and prodigious output, living as he did in a world rich in illusions that protected him from the miseries of his daily existence at the same time as they nourished his flaws. Today, he probably would have been a candidate for lithium therapy, which would calm him, remove some of his drive and energy, and make him a sounder investigator, with perhaps an unpredictable effect on his creative powers. He undoubtedly would not be the Rafinesque we know. Rafinesque's detractors have simply dismissed him as a "mental case," as insignificant, but this is far too simplistic (and hostile) a judgment. It does not take into account his wondrous knowledge and his astounding production in so many fields. Despite all the ups and downs in his life and in the quality and veracity of his work, one can only conclude that he was a brilliant, creative, imaginative man—this is the judgment that remains.

His tragedy was that he had not a shred of insight into why he and his creations were often unacceptable. He was a man who meant every word he said and was passionately convinced of its truth, however absurd or patently in error it was; if he duped anyone it was himself. He was not a conventional charlatan or fraud—rational, deliberately anti-social or criminal, cynically duping his victims. What confounded people in his day, as it does in ours, is that on the one hand he appeared to be a serious scholar of immense learning, who made impressive advances in knowledge, and on the other, he was capable of writing vast amounts of rubbish. Rafinesque's earnest discussion of sea monsters, his ridiculous histories of early North Americans, his astounding financial offer to Poinsett and the Mexican government, his fantastic "inventions" (too amazing to be revealed to the public), his claim to having classified fish as he swam for his life from a shipwreck, and his other extravagant proclamations confuse us because they are intermixed with rational, incisive pronouncements in natural history where he was tethered to facts that could be verified. Unfortunately, too often the reliable could not be separated from the misbegotten, and so in his time, the accumulation of violations resulted in total, outright rejection. Rafinesque differs from William Chester Minor, the lexicographer described in the *Professor and the Madman,* a certified, dangerous schizophrenic whose valuable contributions to the Oxford Old English Dictionary were beyond reproach. Here, mental derangement and professional competence were completely separated.[14]

But beyond the turmoil he created, there are genuinely disturbing and damaging accusations. Some of Rafinesque's colleagues expressed doubts about his honesty; an acquaintance from Kentucky, Charles Wilkins Short, wrote his friend Asa Gray "Everybody knows that poor Raffy was a most bare-faced liar, not to say rogue."[15] There is now very good reason to believe that he *fabricated* important data and documents. (He did not just cut corners, or take liberties.) The most egregious example is the Lenni Lenape migration saga, *Walam Olum,* which has perplexed scholars for one and a half centuries. Rafinesque wrote the *Walam Olum* believing it to be *authentic* because it accorded with his own belief—he was merely recording and giving substance to what *must be* true. It was a damaging, culpably dishonest act, which misled scholars in search of the real truth, far more damaging than his childish creations, which could be easily dismissed; this was more than mischief.[16]

Scientists must have preconceptions about the real world they investigate, but these may change after acquisition of new, verifiable information, reflection, and subjection to the test of logic and common sense. Rafinesque was often unresponsive to these monitoring devices, especially if he deemed the information or insights *new* and *his own*. He was particularly susceptible to seizing upon the *new*, perhaps because he saw in it the possibility of making a name for himself, something he ardently craved.

Rafinesque's authority resided in his head as he attempted to mold nature to the image of his beliefs and impress his conception of order on chaos, only to create more confusion. When he went beyond the data, the authenticity of Rafinesque's empires of the imagination were to him beyond question, rational and grounded in *his* internally verifiable facts drawn from his remarkable archival memory. He was a kind of Baron von Munchausen telling tall tales with absolute conviction about their veracity. When challenged, he reacted with anger rather than reexamining his untenable statements, errors, and distortions—behavior that was interpreted as arrogant, that characterized a man with delusions of grandeur—a judgment that was frequently reinforced by his assertion that what others had done, he could do better. At the same time he was usually generous and magnanimous in giving full credit to his sources of information and specimens.

When Audubon handed him descriptions of several fictitious fish, a seemingly critical mind would have easily discovered the hoax, but Rafinesque, an expert on the subject, "accepted" the information as valid and transferred it to his splendid work on the fish of the Ohio River.[17] The false information, recognized by all as preposterous, should have outraged Rafinesque, but inappropriately it did not, and incredibly, in no way did it harm his friendship with Audubon. Regardless of what anyone said about the data, Rafinesque now believed it and therefore it must be true, and no one could convince him otherwise; they had become *his facts.* Audubon himself must have been surprised by Rafinesque's unquestioning belief—the affair was a practical joke that had gone awry.

In the annals of science, someone like Rafinesque is rare, a person with an affective disorder functioning over a long professional life. It would seem that

Rafinesque's form of mood disorder would be virtually incompatible with the practice of creative science, where data must be reliable and must be properly evaluated, tasks that a scientist on a "high," without the informal restraint of colleagues, cannot accomplish appropriately, thus tainting his work and his reputation. Of course, sound judgments can be made and creditable science accomplished at times when mood swings are stabilized. Overall, this kind of affective disorder is a negative selection factor in science where the scientist is confined by precise observations, *facts,* and *numbers* that allow little or no leeway. By contrast, the prevalence of affective disorders among artists and writers suggests that the condition might be a positive selection factor for the artistically inclined where the restraints of the real world on the artist are undefined and the perception of reality is validly open to endless interpretation. In this regard, Rafinesque occupied a rare, precarious borderland. If indeed Rafinesque suffered a kind of insanity, his affliction should help us evaluate his achievement and understand the basis of his creative genius.

There is so much to extol and so much to despair of in Rafinesque—a man of manifest contradictions who fascinates. Perhaps it is his ferocious passion about nature and the world about him that intrigues us, a passion few can match, leaving us envious of his dangerous gift. Proudly stumbling through life, his manic drive and blithe confidence sustained him, whatever the obstacles. He burned to be recognized, to uncover the great laws of the universe, and to set the world aright, and if he were not heard then, ultimately his views would prevail. What would have crushed ordinary mortals, to the astonishment of his colleagues, he could rise above. Always looking forward to the next challenge, he did not dwell on the litter he left around him. In the history of science he remains memorable, and perhaps unique, not so much for his scientific contributions, which tended not to have a lasting impact, but for the fantastical person that he was.

NOTES

Preface

1. The discovery and history of the manuscript of *A Life of Travels,* is presented in the Preface to the original French version (1987).

2. Recently, *Proflies of Rafinesque,* a collection of informative essays on Rafinesque, many previously unpublished, by multiple authors, edited by Charles Boewe and published by The University of Tennessee Press, Knoxville (2003).

Introduction

1. Rafinesque, *New Flora,* part 1, pp 11–15.

Chapter 1. In the Beginning

1. Most of the details in this chapter and the narrative elements throughout the book have been taken from Rafinesque's short autobiography, *Life of Travels,* written in French in 1833. It was republished in 1944 in *Chronica Botanica,* vol. 8, pp. 293–353. Fitzpatrick (1911), Pennell(1942), and Boewe (1982) are also excellent sources of information about Rafinesque and his publications. Throughout the notes, page numbers of *Life of Travels* refer to those of *Chronica Botanica;* quotes without a number also refer to this work.

2. The surname Schmaltz was added to the father's, according to Latin tradition. In this instance the exquisite *Rafinesque* was lengthened to *Rafinesque-Schmaltz,* according to the political exigencies of the time. Occasionally, Constantine was referred to as Schmaltz-Rafinesque, or Constantine Schmaltz (instead of Samuel) Rafinesque. In any case, his noble first and Old Testament middle names and the euphonious *Rafinesque* were bludgeoned by the name of his mother.

3. *Travels,* p. 299.

4. Ibid., p. 353.

5. Ibid., p. 300.

6. Richardson, 1982; Curti et al, pp. 217–301.

7. Joseph Leidy to Augustus A. Gould, May 11, 1848, BPL.

8. See Bruce, 1987, pp. 65–68.

9. De Crevecoeur, letter 3, *What is an American?* pp. 66–105.

Chapter 2. The Newcomer in Botanical Paradise, 1802–1805

1. The quote by Rafinesque is found in the *Cincinnati Literary Gazette*, vol. 2, 1824, p. 204.

2. *Travels,* p. 301.

3. Richardson, 1982, pp. 208–57; Smith, 1995, pp. 10–14.

4. See Richmond, 1909, pp. 249–51. These papers were sent to Daudin who had them published in the *Bulletin of the Société Philomathique* of Paris, 1802.

5. Jefferson to Caspar Wistar, June 21, 1807.

6. Excellent accounts of science and botany in the Philadelphia area can be found in Greene, 1984; Porter, 1986; Barnhart, 1909; Stuckey, 1971.

7. Specimens of this tree can still be purchased from the Bartram gardens.

8. De Crèvecoeur, letter 3, *History of Andrew, the Hebridean,* in *Letters,* pp. 90–105.

9. Rafinesque, *A Life of Travels,* pp. 302–3; Smyth; throughout the text, all unattributed quotes can be found in Rafinesque's *A Life of Travels ("Travels").*

10. *Travels,* p. 303.

11. Ibid., pp. 302–4.

12. Betts, E.M., 1944.

13. Rafinesque to Jefferson, Nov. 27, 1804; letters published by Betts, 1944.

14. Jefferson to Rafinesque, Dec. 15, 1804.

15. Thomas Clifford to Jefferson, Dec. 18, 1804.

16. It is not known whether they were separate manuscripts.

17. *Philadelphia Medical and Physical Journal,* under Miscellaneous Facts and Observations, Botany, 1805: "Mr. Rafinesque, an Italian gentleman, has put into the hands of the Editor, a MS. catalogue of the plants of the State of Delaware and the District of Columbia. This catalogue, [with large additions by the Editor], will be published in future parts of this journal."

18. The Barton-Rafinesque affair is discussed in detail in Little, 1943; Barton's relationship with his friend William Bartram is briefly discussed in Slaughter, pp. 229–30; also see Barton letter to W. Bartram, August 26, 1787, HSP.

19. Latin: *make haste slowly*; Muhlenberg, H., to Stephen Elliot, June 16, 1809; Merrill, E. D. 1948, p. 6. The letter is in the Gray Herbarium Library of Harvard University.

20. Stuckey, 1971; also see Gordon, R. B., *Science,* vol. 115, 1952, pp. 217–18.

21. See Lindroth, S. on Linnaeus in DAB, vol. 8, 1973, pp. 374–81; Stearn, 1959.

22. Ernst Mayr, *Systematic Zoology,* vol. 3, 1954, pp. 86–89.

23. Stearn, 1959, p. 5.

24. Antoine-Laurent de Jussieu, *Genera plantarum,* Paris, 1789; See Stafleu, F.A. on de Jussieu, A-L, in DSB, vol. 7, 1973, pp. 198–99.

25. Rafinesque, *Annals of Nature,* 1820, p. 71.

26. Merrill, 1949, p. 8; Rafinesque, *Alsographia Americana,* pp. 3–4.

27. Simpson, 1990, pp. 107–46; Fulling, E.H. 1945, pp. 249–51.

28. Joseph Francis Correa da Serra (1750–1823), a member of the ANS.

29. Rafinesque, *Saturday Evening Post,* vol. 6, March 10, 1827.

30. Cain, pp. 13–23; Rafinesque expounded his views on genus and species in his *Flora Telluriana,* (first part, 1836), pp. 90–101.

31. Rafinesque, *Flora Telluriana*, part 1, 1836, pp. 16–17; Ibid., *Amenities of Nature*, 1840; see Quatre, H., Archivio Botanico, 3rd series, vol. 24, 1948, p. 182.

32. Rafinesque frankly stated his opinions in the form of a letter to Bory St. Vincent in his publication *Western Minerva,* Lexington, Kentucky, January 1821, pp. 72–73.

33. Stevens, 1998.

34. The problem of defining the boundaries of species and the permissible limits of variation within a species is very much alive today. Mark Cocker (*Manchester Guardian Weekly*, September 16–22, 1999, p. 20) has written about the controversy concerning the taxonomy of the common herring gull. A new scheme of classification may create eight species where only one formerly existed, based on characteristics that were formerly ignored, such as the color of legs (pink versus bright yellow), and peculiarities that have arisen in gulls in different parts of the world. A partial solution has been to subdivide species into races and subspecies, but according to Cocker "this has been shown to be extremely arbitrary and serves little more than an *administrative convenience*." According to "a new approach" in systematics known as *phylogenetic species concept*, which sounds much like Rafinesque's species splitting philosophy that even embraces his notions about transformism, a true species consists of a cluster of individuals with "essential and irreducible" characteristics arrived at in the course of evolution. Adoption of this system would more closely align taxonomy with the evolutionary process, but a consequence would be that the number of species of birds would suddenly rise from 10,000 to 25,000, a change that, according to conventional ornithologists, would create turmoil and discourage wide interest in birds because of technical complexities. Some of these arguments were used against Rafinesque.

35. Simpson, G.G., 1990, pp. 107–46; pp. 171–86.

36. E.D. Merrill, 1948, p. 7.

37. Ibid., 1946; Cain, pp. 18–20.

38. Rafinesque, "Monograph of the *Bivalve Shells of the Ohio and Other Rivers of the Western States,* Brussels, 1820"; republished with addendum in 1831.

39. Ibid.

40. Porter, 1979.

41. Daniels, 1967; see D.E. Allens enlightening social historical discussion of natural history and science.

42. Ord, 1834, pp. xii–xvi. Ord was critical of Thomas Say's classic *American Conchology* despite the fact that they were both members of the Academy of Natural Science's expedition to eastern Florida in 1817 and had published papers together.

43. Stroud, 1992, pp. 52–54.

44. Evans, pp. 15, 27, 86, 155.

45. For the early ideas about evolution and the origin of species held by Darwin see Darwin, 1909; Mayr, pp. 1–9.

46. Rafinesque, "Principle of the Philosophy of New Genera and New Species of Plants and Animals," *Atlantic Journal, and Friend of Knowledge,* No. 124, pp. 163–64, 1833. This article was copied from Rafinesque's letter to John Torrey, Dec. 1, 1832.

47. Darwin, in his *On the Origin of Species by Means of Natural Selection,* 6th edition, 1901, p. xviii, quotes Rafinesque's comments in *New Flora of North America,* (1836).

48. M. Adanson, *Familles des Plantes,* vol. 1, 1763.

49. Rafinesque, *The World, or Instability,* 1836, line 1282, and note 22, p. 222.

50. Jordan, 1888, p. 154.

51. Gray, 1841.

52. Rafinesque became a naturalized citizen in 1832.

Chapter 3. Sicily, 1805–1815

1. The narrative components and the quotations of this chapter are taken from Rafinesque's autobiography, *A Life of Travels. . .* unless otherwise indicated.

2. *Travels,* pp. 306–7.

3. Pennell, 1942.

4. An extremely useful, complete bibliography of Rafinesque's writings is found in Boewe (1982), pp. 83–260. The book also contains *Rafinesque, A Sketch of His Life With Bibliography,* by T.J. Fitzpatrick, 1911. The Rafinesque papers in *The Medical Repository* are listed in Boewe as nos. 5(1806), 6, 8, 9, 10, 19, 21, 22(1811), all but 1 on North American plants, pp. 85–86. The papers were later translated into French and republished in Desvaux's *Journal de Botanique,* in Paris. Boewe's book also contains a chapter on the eccentric bibliophile, T.J. Fitzpatrick, an expert on Rafinesque, *Fitzpatrick, Man and Bibliographer,* pp. 1–18.

5. Rafinesque's zoological papers, including those from Sicily, were critically discussed by S.S. Haldeman in the *American Journal of Science and Arts,* vol. 42, 1842, pp. 280–91.

6. Asa Gray, *American Journal of Science and Arts,* vol. 40, 1841, pp. 221–41.

7. Isaac Lea, *Proc ANS,* vol. 7, 1854, pp. 236–49.

8. Swainson, pp. 300–301.

9. Rafinesque, *Life of Travels.,* p. 312.

10. Swainson to Harlan, Dec 7, 1828. Gray Herbarium Library, Harvard University.

11. The library of the Linnean Society holds 52 of these letters. They are listed in the Proceedings of the Linnean Society of London, Session 1899–1900, pp. 49–51.

12. Gillispie, pp. 44–45; Doskey, pp. xxiv–xxvi.

13. *Travels,* p. 310.

14. Ibid., p. 311.

15. Rafinesque Schmaltz and G. E. Ortolani, *Statistica generale di sicilia,* 1810.

16. See Cain, 1990, for translations of Rafineque's Sicilian publications and for commentary. Quotes are from Cain's publication.

17. *Travels,* pp. 313–14.

18. Weiss, 1936, p. 30.

19. Call, 1895.

20. Rafinesque, *Epitome . . .* in Cain, 1990, p. 62.

Chapter 4. American Science and Natural History in Rafinesque's Time

1. See Goode, pp. 109–10.
2. Daniels, 1968, chapter 2, p. 34; Ibid., 1967, chapters 1–3.
3. See Nathan Reingold in Daniels, 1972; John C. Greene, *Journal of American History*, vol. 55, 1968, pp. 22–41.
4. George Ord, *A Memoir of Thomas Say . . .* , read before the APS on December 19, 1834. The acerbic Ord berated Say for devoting his entire life to science in the socialistic New Harmony community, of which he disapproved.
5. Keeney, 1992, discusses the amateur botanist in the United States.
6. Betts, p. 373. Five of eight professors hired were of European origin.
7. Johns, 1994.
8. Cheyney, p. 205; in the late twentieth century, support for the two schools was intimately coupled because the value of basic research in the understanding, treatment, and cure of diseases is fully appreciated. This coupling did not exist in the nineteenth century, so there was no need to support basic research for its own sake. Today, support for basic research is relatively generous because of the promise of elimination of the ailments of humankind.
9. Trollope, 1832.
10. Dickens, 1843; 1842.
11. Samuel Johnson, *The Journal of a Tour to the Hebrides.*
12. Rusk, R.L., 1925, vol. 1, pp. 1–5.
13. Sheridan, pp. 39–42.
14. Dickens, p. 547.
15. Baatz, 1986; De Buffon was probably influenced by his certainty that the produce and cuisine of France were superior to that of America and that the conditions of his life in France were gentler than those of his American counterparts.
16. A brilliant and highly informative discussion of this matter can be found in Gerbi.
17. See Kohlstedt, 1991, pp. 108–9.
18. Porter, 1979a; 1986, chapter 4: Graustein, 1967, pp. 82–93.
19. Morton, 1841; Stroud, pp. 32–34.
20. William Swainson to Richard Harlan, December 7, 1828. Herbarium Library, Harvard University.

Chapter 5. Return to America, 1815–1818

1. *Travels,* p. 314. Much of the narrative part of the chapter is taken from Rafinesque's autobiography from Fitzpatrick, 1911 and Boewe, 1982.
2. *Travels,* p 315.
3. Rafinesque's letter was seen by Van Welden and published in 1822 in the Ger-

man journal *Flora.*, recounted by Merrill, 1948, p. 19. See *American Monthly Magazine and Critical Review*, vol. 2, 1818, pp. 243–44.

4. *Travels,* p. 315.

5. Some of the important catalogues of American plants published in the early nineteenth century were: *North American Flora* by André Michaux, (1803), the *History of the Forest Trees of North America* by his son François-André Michaux, (1818–1819) (American Edition), and others by Henry Muhlenberg (1813), Frederick Pursh, Jacob Bigelow (1814), Thomas Nuttall(1818), Stephen Elliot (1816–1824), and William P. C. Barton, (1818, 1820–1824).

6. Rafinesque, *Circular Address on Botany and Zoology* (on Microcard).

7. Rodgers, 1942, chapters 2 and 3. Rafinesque in Palermo to Manasseh Cutler near Salem, Mass., May 2, 1806. (An unpublished Letter by Rafinesque, R.E. Call, *Science*, vol. 15, pp. 713–14, 1902.)

8. *American Monthly Magazine and Critical Review*, vol. 4, 1819, pp. 465–67.

9. Baatz, 1990, pp. 22, 32–34.

10. Swainson to Rafinesque, June, 1820, in Betts, p. 375.

11. James, J.T. A letter from Torrey to Eaton, *Botanical Gazette*, vol. 8, 1883, pp. 289–91.

12. Eaton to Torrey, Oct. 5, 1817.

13. Ibid., March 24, 1818.

14. Rafinesque's introduction to *Florula Ludoviciana,* C. Wiley and Co., NY, 1817.

15. DeKay, p. 31.

16. Torrey to Eaton, March 21, 1816.

17. Ibid., Jan. 5, 1818.

18. Gray, 1841, p. 233.

19. Pennell, 1940.

20. Ibid., 1942, pp. 17–18.

21. Shinners, Lloyd H., *Rhodora,* vol. 59, pp. 265–67, 1957.

22. Cain, 1990.

23. Rafinesque, *American Monthly Magazine and Critical Review,* "Descriptions of seven new genera of North American Quadrupeds," vol. 2, 1817, pp. 44–46; *Atlantic Journal and Friends of Knowledge,* "Six new Firs from Oregon," pp. 119–20.

24. Eaton to Torrey, April 16, 1818.

25. Eaton in *The Troy Whig,* Jan. 20, 1835.

26. Merrill, 1942. Merrill, a history-minded botanist, was one of the major champions of Rafinesque. In this paper and in his 1943 and 1949 publications, he discussed the basis for the turmoil Rafinesque created in American and world botany. The present discussion is based on these papers and that of Pennell, 1942, and Fernald, M.L., *Rhodora,* vol. 34, 1932, pp. 21–28; vol. 44, 1944, pp. 1–21, 32–57, and 310–11.

27. Merrill, 1942, p. 79.

28. Stuckey, 1971a.

29. Gray, 1841, p. 234.

30. Rafinesque to James Mease, January 1, 1816. Document 816, University of Pennsylvania Archives.

31. Mitchell, document 1951, University of Pennsylvania Archives.

32. Cheyney, 1940, pp. 202–4.

33. Betts, 1944; contains all of the correspondence of Jefferson and Rafinesque.

34. All articles of *American Monthly Magazine and Critical Review* are on Microfilm (APS reel 58).

35. Rafinesque, *American Monthly Magazine and Critical Review*, vol. 2, 1817, pp. 81–89.

36. Ibid., vol. 1, no. 6, 1818, pp. 426–30. Rafinesque took offense at being called an "Italian naturalist" by Eaton. In his third person response to Eaton's remarks Rafinesque wrote, "He gives to Mr. Rafinesque the title of an Italian naturalist. Of the title of a naturalist we believe Mr. R. will always be proud, but he never dreamt of being an Italian, any more than the American citizen who travels, resides in various parts of Italy!"

37. Maclure, *Observations on the Geology of the United States of America*. This was the first comprehensive geological work in the United States, and Maclure is considered by many to be the "father" of American geology.

38. Rafinesque, *American Monthly Magazine and Critical Review*, vol. 1, 1817, pp. 356–59.

39. Ibid., vol. 2, 1817, pp. 143.

40. Ibid., vol. 2, pp. 265–69.

41. Ewan, p. 628.

42. Rafineque, *American Monthly Magazine and Critical Review*, vol. 2, 1818, pp. 202–3.

43. Ibid., vol. 3, 1818, pp. 269–74; Rafinesque to Short, September 27, 1818, in Perkins, p. 220.

44. Rafinesque to Ord, Oct 1, 1817 (ANS).

45. Rafinesque, *American Monthly Magazine and Critical Review*, vol. 4, 1819, pp. 184–96.

46. Rafinesque to Torrey, Feb 1, 1821, HSP.

47. Rafinesque, *American Monthly Magazine and Critical Review*, vol. 1, 1817, p. 274

48. L. Agassiz, *American Journal of Science and Arts*, vol. 17, 1854, pp. 297–308, and 353–65.

49. Rafinesque, *American Monthly Magazine and Critical Review*, vol. 4, 1819, pp. 382–84.

50. Ibid., *Atlantic Journal and Friend of Knowledge*, vol. 1, 1833, pp. 200–201.

51. Ibid., *Western Minerva*, vol. 1, 1820, pp. 43–46.

52. Ibid., pp. 35–37.

53. Ibid., *The Western Review and Miscellaneous Magazine*, vol. 1, 1819, pp. 185–86; *Atlantic Journal and Friend of Knowledge*, vol. 1, 1832, pp. 12–13; Ibid., (sulphur), pp. 13–14.

54. Ibid., *Atlantic Journal*, pp. 32–34.

55. Hance, Anthony M., *Rafinesque, the Great Naturalist*, A paper read before the Bucks County Historical Society at Langhorne, June 4, 1914.

56. Torrey to Eaton, April 16, 1818, N.Y. State Library, Albany, New York.

Chapter 6. Going West, 1818

1. See Boewe, 1988: Rafinesque, *Life of Travels*, 1836, pp. 53–58; Pennel, 1942, pp. 18–23. Rafinesque's travels are described in four letters to Zacheus Collins, friend and patron of Rafinesque, vice-president of the ANSP, with botanical interests. The narrative parts of this chapter are from Rafinesque's *Life of Travels*.

2. Audubon, vol. 1, 1831, pp. 455–60; Audubon, 1926, pp. 97–104; see also Herrick, 1917, pp. 285–300.

3. Jordan, 1888.

4. Swainson, p. 301.

5. Binney, pp. 38–55.

6. Call, pp. 24–36.

7. The hoax was clarified by David Starr Jordan (1886); A. Wheeler, *Bulletin of Zoological Nomenclature,* vol. 45, 1988, pp. 6–12; Markle, 1997 for details of the spurious fish. Of the fourteen fish, eleven were derived from Audubon. Rafinesque was careful to name his sources of information.

8. An explanation for Rafinesque's lack of resentment, proposed by Boewe (personal communication) is that the "M. de T." Audubon speaks of might not have been Rafinesque but another Frenchman a doubtful possibility because no other human could fit Audubon's description.

9. David S. Jordan, 1886.

10. Pennell, 1942, pp. 19–23. Rafinesque's trip was described in four letters to Zaccheus Collins. The letters are summarized by Pennell.

11. Jordan and Butler, 1927; Peattie, 1936, pp. 251–63; Arndt, 1965.

12. Rafinesque to Charles Wilkins Short, Dec. 21, 1819; reprinted in Perkins, 1938.

13. Ibid., Sept 27, 1818, cited in Perkins, 1938, p. 203.

14. Clifford's articles were published as letters, which in 2000 were assembled by Boewe; (see Clifford, 2000, in the bibliography).

15. A lucid account of Rafinesque's archaeological interests has been written by Stephen Williams, 1991, pp. 98–115.

16. *Travels,* p. 319.

17. Rafinesque, *American Monthly Magazine and Critical Review,* vol. 4, 1818, pp. 106–7.

Chapter 7. Kentucky, 1819–1826

1. The narrative and unattributed quotes are from *Travels.*

2. Pierson, 1959, pp. 122–26.

3. Four informative books on Transylvania University and Lexington are: Caldwell, 1828; Jennings, 1955; Wright, 1975; Gobar and Hamon, 1983. *Transylvania* was a name given by Virginians of classical bent for an area of Kentucky that lay "beyond the forest," just as the Romans had called Eastern Europe *Transylvania.*

4. Call, 1895, p. 32; Dupre, 1961; Greene, pp. 120–27.

5. Dupre, p. 111; for an excellent discussion of the early years of Lexington, Pittsburgh, Louisville, and other near-western cities see Wade, 1964.

6. Ida Withers Harris, *Journal of American History,* vol. 7, 1913, pp. 901–9; Wright, pp. 82–83.

7. Wright, p. 114.

8. Rafinesque, in *Western Minerva,* Fragments of letters from Lexington, to a Lady, p. 76.

9. Jordan to Mrs. Norton, April 23, 1924, mentioned in Wright, pp. 354–55.

10. Despite the mildness of his sympathy for incorporating scientific subjects

into the curriculum, he made an effort to hire Benjamin Silliman of Yale as a professor of chemistry; the effort was unsuccessful.

11. Call, 1895, p. 42. Information requested by Dr. Call from a surviving member of the Holley household.

12. Lee, p. 168. Rafinesque is frequently mentioned in Mary Austin Holley's biography; also see Call, p. 43, 63.

13. Jennings, p. 121.

14. Boewe, 1983, p. iii–iv.

15. Archives of the ANS.

16. Courses outlined in *Prospectus of Two Courses of Lectures on Natural History and Botany,* ANS.

17. Weiss, 1936, Introduction; Jennings, pp. 104–5, 122.

18. *Kentucky Reporter,* January 12, 1824. Cited in Boewe, 1983.

19. Call, 1895.

20. Peattie, p. 265, taken from Fitzpatrick, 1911.

21. Funkhouser, pp. 13–14.

22. Rafinesque, "Thoughts on Atmospheric Dust," *Silliman's Journal,* vol. 1, 1819, pp. 397–400; Rafinesque's theory was disputed by "XYZ" in the same journal, vol. 2, 1819, pp. 134–36.

23. Rafinesque, *American Monthly Magazine and Critical Review,* vol. 2, 1818, pp. 39–42.

24. Ibid., *Kentucky Reporter,* March 15, 1820. Cited in Boewe, 1983.

25. Ibid., *Atlantic Journal and Friend of Knowledge,* vol. 1, 1833, pp. 167–70.

26. Weiss, 1936, pp. 60–61.

27. Rafinesque, *Celestial Wonders and Philosophy,* Philadelphia, 1838.

28. A list of Rafinesque's articles in *The Western Review and Miscellaneous Magazine* is presented by Boewe, 1982, based on compilations by Fitzpatrick, 1911.

29. Rafinesque, *Travels,* p. 320.

30. Ibid., *Western Minerva,* 1820, p. 49.

31. Ibid., pp. 64–65.

32. Ibid., *Travels,* p. 322; Call, 1895, pp. 169–70. An original copy of the journal can be found in the archives of the ANS.

33. Ibid., *Travels,* p. 322.

34. Rafinesque to John Torrey, Feb. 1, 1821, HSP.

35. Rafinesque, *Western Minerva,* 1820, p. 74.

36. Ibid., *Ancient History or Annals of Kentucky with a Survey of the Ancient Monuments of North America,* printed by the author, Frankfort, Kentucky, 1824.

37. Ibid., *Cincinnati Literary Gazette,* vol. 1, 1824, pp. 59–60.

38. Ibid., vol. 2, 1824, pp. 50–51.

39. Ibid., pp. 116–17.

40. Ibid., p. 70.

41. Ibid., vol. 1, 1824, p. 70.

42. Ibid., vol. 3, 1825, pp. 89–90.

43. Boewe, 1982, p. 143, reference 449.

44. Rafinesque, *Cincinnati Literary Gazette,* vol. 1, 1824, pp. 161–62.

45. Ibid., vol. 3, 1825, pp. 66–67. The plea to reveal Rafinesque's secrets appears on p. 90 of the *Gazette.*

46. Rafinesque, *American Monthly Magazine and Critical Review*, vol. 3, 1818, pp. 41–44.

47. Ibid., *Travels*, p. 330.

48. The archives of the Linnean Society (London, U.K.) possess fifty-two letters from Rafinesque to Swainson, the first dated May 25, 1809, and the last on April 12, 1840, just five months before he died.

49. Cuvier to Rafinesque, April 28, 1828, in Betts, p. 375.

50. See Perkins 1938 for the entire series of letters, with the commentary of Perkins.

51. Rafinesque to Short, December 21, 1819, in Perkins, 1938, pp. 222–24.

52. Ida Withers Harris, *Journal of American History*, vol. 7, 1913, pp. 901–9; Gobar and Hamon, pp. 37–38.

53. Rafinesque to Zaccheus Collins, cited in Pennell, pp. 28–30.

54. Rafinesque, *Travels*, p. 324.

55. Ibid., *Medical Flora or Manual of the Medical Botany of the United States of North America*, 1828.

56. Rafinesque to Jefferson, Feb. 15, 1824. The entire correspondence is found in Betts, 1944.

57. Ibid., Sept. 16, 1819.

58. Jefferson to Rafinesque, Nov. 7, 1819.

59. Rafinesque to Jefferson, Jan. 25, 1821.

60. Ibid., July 1, 1824.

61. Jefferson to Rafinesque, Aug. 11, 1824 (Betts, 1944).

62. Rafinesque to I. Wetmore, University of North Carolina, April 22, 1826.

63. Perkins, 1938, pp. 209–211; Rafinesque to Short, Feb 1, 1822.

64. Daniels, 1967; see Nathan Reingold in Daniels, 1972; John C. Greene, *Journal of American History*, vol. 55, 1968, pp. 22–41.

Chapter 8. The World of Finance and Banking

1. Rafinesque, *Travels*, p. 340. Numerous misspellings of proper names may suggest a deterioration of mental function in the last decade of Rafinesque's life.

2. Rafinesque to Collins, June 16, 1825 (HSP).

3. Johnson, pp. 851–60.

4. Fitzpatrick, 1911, p. 32.

5. Rafinesque, *Safe Banking*, p. 58.

6. Ibid., p. 88.

7. Rafinesque, *The Pleasures and Duties of Wealth*, Philadelphia, 1840, pp. 21–27.

8. Christopher C. Graham to Clay, August 31, 1825, in Hopkins, vol. 4, 1959, p. 609.

9. Rafinesque, *American Manual of the Grape Vines*, 1830.

10. Rafinesque to Collins, Oct. 10, 1825 (HSP).

11. Ibid., Oct. 11, 1825.

12. Rafinesque, *Travels*, p. 330.

13. Ibid., p. 76.

14. Ibid., p. 76, 87.

15. Rafinesque to Poinsett, May 5, 1825, HSP.

Chapter 9. Travels and Farewell to Lexington

1. Many of Rafinesque's notebooks are in the National Library, Smithsonian Institute.

2. Rafinesque to Collins, Nov. 22, 1825; Jan. 12, 1826.

3. Wright, pp. 99–116.

4. Dupre, 1961, pp. 117–18; Gobar and Hamon, pp. 38–45.

5. Rafinesque, *Travels*, p. 327.

6. Porter, 1986, pp. 87–121.

7. *New Harmony Gazette*, May 17, 1826, pp. 269–70.

8. The name was later changed to Rensselaer Polytechnic Institute.

9. Rezneck, 1959.

10. Reingold, pp.152–53. Henry recounts Rafinesque's "very important observation" that poisonous snakes have broad jaws, housing the fangs [and glands], making them wider than the snake's neck.

11. Eaton to Anna B. Eaton, April 3, 1826.

12. Eaton in *The Troy Whig*, Jan. 20, 1835; see McCallister, for interactions between Rafinesque and Torrey, pp. 222–23, 253–57, and 480.

13. Boewe, 1962, pp. 58–60. The name Mt. Rafinesque became official some years later.

14. Rafinesque to Eaton, July 30, 1826 (Syracuse University Library). The letter is of peculiar interest because one of the paragraphs is written in an entirely different style.

15. Rafinesque to Torrey, Jan. 2, 1832.

16. Rafinesque to Collins, Aug. 14, 1826.

17. Collins to Rafinesque, Aug. 16, 1826.

Chapter 10. The Medicine Man

1. Jefferson to Caspar Wistar, June 21, 1807. The discussion of early nineteenth-century medicine, especially of the botanical variety, was largely derived from Rothstein (1988), Wallace (1980), Berman (1952, 1956), Shryock (1960), and Weeks (1945).

2. See Myer, p. 1–21.

3. See Shryock, p. 1–38.

4. For a brief but informative discussion of home medicine, see Charles C. Rosenberg in *Every Man His Own Doctor: Popular Medicine in Early America*, The Library Company of Philadelphia, 1998.

5. These quotes were taken from Meyer, p.11.

6. Rothstein, pp. 42–43; Thomson, 1819, 12 ed., 1841. Thomson was a prolific writer who frequently published the texts of his speeches made while traveling about the country visiting Friendly Societies.

7. Rafinesque, vol. 1, 1828, p. iv.

8. Barkley, p. 72; Webster, 1898.

9. Berman, 1952, pp. 414–15.

10. Miller, pp. 24–57; The Shaker catalog carried the endorsement of "Dr. Rafinesque": "The best medical gardens in the United States are those established by

the communities of the Shakers, or modern Essenians, who cultivate and collect a great variety of medical plants. They sell them cheap, fresh and genuine," p. 31; see also Berman, 1960, which contains a letter from Lawrence to Rafinesque.

11. Rafinesque, *Saturday Evening Post*, June 16, 1827; see *Travels*, p. 330.

12. Ibid., articles and advertisements in issues of Oct. 6, Nov. 17, and Dec. 29, 1827; see Weeks, 1945.

13. Weaks, 1945; Merrill, 1949, p. 39.

14. Drake, 1830.

Chapter 11. History, Archaeology, and Linguistics

1. Thomas, pp. 17–50.

2. Rafinesque, *Saturday Evening Post*, vol. 5, number 44, 1826, p.1; Red man, Ibid., June 7, 1828, p. 1.

3. Ibid., *Atlantic Journal and Friend of Knowledge*, vol. 1, 1832, pp. 10–11.

4. Samuel George Morton, *Crania Americana, or a Comparative View of the Skulls of Various Aboriginal Nations of North and South America*, Philadelphia, 1839.

5. Rafinesque, *Cincinnati Literary Gazette*, vol. 2, 1824, pp. 203.

6. Ibid., p. 178.

7. Ibid., *Saturday Evening Post*, vol. 7, 1828, p. 1.

8. Ibid., *The American Nations*, 1836, Introduction.

9. Ibid., *Atlantic Journal and Friend of Knowledge*, vol. 1, 1832, pp. 85–86; Ibid., Reward of Merit, *Atlantic Journal and Friend of Knowledge*, vol. 1, 1833, p. 157. The medal is now in the possession of the College of Physicians of Philadelphia.

10. Ibid., pp. 198–99.

11. Ibid., *Saturday Evening Post*, April 14, 1829, p. 1.

12. Ibid., vol. 7, June 7, 1828, p. 1.

13. Jefferson, chapter 11.

14. Stoltman, 1973, pp. 116–50.

15. Chapter 14, pp. 376–408, of Greene, provided much of the background for the discussion on linguistics and the origin of the Native American.

16. Jefferson, *Notes on the State of Virginia*.

17. B.S. Barton, *New Views of the Origin of the Tribes and Nations of America*, 2nd ed., Philadelphia, 1798.

18. For an excellent review of the early archaeology (and of eighteenth-century science in general) of the United States, see Greene, chapter 13, pp. 343–75.

19. Rafinesque, *Atlantic Journal and Friend of Knowledge*, vol. 1, 1831, pp. 48–51.

20. Rafinesque, *Historical and ethnographical Palingenesy* in *The Good Book, and Amenities of Nature*, 1840, pp. 68–70.

21. Belyi, p. 65; Rafinesque, *Atlantic Journal and Friend of Knowledge*, vol. 1, 1832, pp. 161–63.

22. Rafinesque, *Cincinnati Literary Gazette*, vol. 1, 1824, pp. 59–60; *Atlantic Journal and Friend of Knowledge*, vol. 1, 1832, pp. 6–8.

23. Ibid., in Priest, pp. 315–22; Ibid., *Atlantic Journal and Friend of Knowledge*, vol. 1, 1832, pp. 40–44.

24. Ibid., *Atlantic Journal and Friend of Knowledge*, vol. 1, 1833, pp. 173–75.

25. Ibid., *Saturday Evening Post*, vol. 6, 1827, p. 2. A second open letter to

Champollion on the relation of the Mayan Glyphs to "Lybyan" and Egyptian (all "Atlantic alphabets") was published in Priest, pp. 123–29.

26. Ibid., vol. 7, June 7, p. 1; Ibid., June 21, p. 1; Ibid., July 19, p. 2; Ibid., Sept. 6, 1828, p. 1.

27. Stewart, pp. 18–19.

28. Rafinesque, *Atlantic Journal and Friend of Knowledge,* January 1832, pp. 4–6; Ibid, pp. 6–8; Ibid, February 1832, pp. 40–44.

29. Boewe, 1985, p. 108.

30. See Priest, pp. 312–52; Williams, p. 53.

31. Rafinesque, *Atlantic Journal and Friend of Knowledge,* vol. 1, 1833, pp. 101–5.

32. Rafinesque, *Western Review and Miscellaneous Magazine,* vol. 1, 1819, pp. 313–14.

33. Ibid., vol. 2, 1820, pp. 242–44; vol. 3, 1820, pp. 53–57.

34. For a review of investigations of early man in America see Greene, chapters 12, 13, and 14; Wauchope has written about the various bizarre theories about the origins of aboriginal Americans (Egyptian, Atlantean, Israelite, etc).

35. Haven, pp. 39–40.

36. Rafinesque, *Clio* no. II; *The Cincinnati Literary Gazette,* vol. 1, 1824, pp. 107–8.

37. Atwater to Parker Cleaveland, cited in Greene, pp. 371–72.

38. Boewe, 1987, pp. 44–49.

39. Rafinesque, *Cincinnati Literary Gazette,* vol. 1, 1824, pp. 59–60.

40. Brinton, p. 150.

41. Rafinesque, *Cincinnati Literary Gazette,* vol. 3, 1825, pp. 89–90.

42. Ibid., vol. 2, 1824, pp. 202–4.

43. Brinton, p. 151.

44. Rafinesque, *Saturday Evening Post,* vol. 7, June 21, 1828, p. 1; vol. 7, Sept. 6, 1828, p. 1. The ignorance of earlier Spanish history is evident in the organizers and participants of the Long expedition to the Rocky Mountains in 1819–1820. After reading Evan's fascinating account and commentary of the heroic efforts of the group, one is struck by how often they had been preceded by the Spanish (and the French), whose early efforts were almost disregarded. Perhaps their accounts were not available (in English) to Long and his associates. In an article by Rafinesque and signed *Historicus,* (Ibid., vol. 7, March 22, 1828), he reviewed the histories of America that had been written, pointing out their inadequacies. To write a proper history one would have to read extensively in at least six languages and carefully examine the ancient monuments. "The attempt has been made by an industrious, persevering and well qualified friend of ours, Professor Rafinesque. It is not without due deliberation that we venture to recommend his AMERICAN HISTORY." He added that Rafinesque had proven "beyond a doubt the former existence of many great Empires on our Soil." Apparently he felt that such thinly veiled praise of himself would deceive his readers. Rafinesque often wrote of himself in the third person, too modest to declare the truth about himself in his own voice.

45. Ibid., vol. 7, March 22, 1828; see Stewart, 1989.

46. Haven, pp. 39–40.

47. Ethan Smith; Priest, pp. 58–82.; Wauchope, pp. 53–68.

48. Rafinesque, *Atlantic Journal and Friend of Knowledge,* vol. 1, 1832, pp. 98–99; see Boewe, (1985), on Rafinesque's confusing involvement with Mormon belief. Clearly,

Rafinesque denied that Jews were in America before Columbus, directly at odds with Mormon mythology.

49. Rafinesque, *Saturday Evening Post,* vol. 8, Sept 12, 1829, p. 1; *Atlantic Journal and Friend of Knowledge,* vol. 1, 1832, pp. 98–99; see Wauchope, pp. 5, 50–68; Rafinesque, 1838, p. 7.

Chapter 12. *Walam Olum*

1. Rafinesque, *American Nations,* 1836; Ibid., in Brasseur de Bourbourg, M., 1868, pp. 435–37; Newcombe, 1955, p. 57.

2. Weslager, chapter 4, 1972, pp. 77–97.

3. Squier, 1849.

4. Hoffman, pp. 9–14; Daniel Hoffman's poem celebrates the history, legends, and lore of the Lenni Lenape Indians. The poem tells of their encounter with a fair and honorable Quaker, William Penn. Later generations of white men (including some of Penn's descendents) violated Penn's treaty with the Indians in word and deed, depriving them of their land and honor; ultimately they were the agents of extinction of the Lenni Lenapes. (In fact, Rafinesque was highly critical of Penn's dealings with the Lenni Lenapes, claiming that he had purchased vast amounts of land for a few baubles.) Part of the *Walam Olum* is incorporated into Hoffman's opus, a work that has been set to music by Ezra Laderman. This exquisite oratorio was premiered in Philadelphia on March 4, 2000. The beauty, poignancy, and noble sentiment of this creation is in no way diminished by the historical inaccuracy of the story, given that recent, authoritative analysis makes it almost certain that the *Walam Olum* is a fraud authored by Rafinesque.

5. McCutcheon, Preface, pp. xi–xiv.

6. Brinton, 1885; Rafinesque's manuscript is in the archives of the Universiy of Pennsylvania.

7. Another heroic but unsuccessful effort to identify Dr. Ward was made by Paul Wier, (Proceedings of the Indiana Academy of Science, vol. 51, 1941, pp. 55–59); see Barlow and Powell (1986), who argue that Malthus A. Ward was the man in question, while Boewe (1987a) disputes the notion.

8. Wier, pp. 158–59.

9. Heckewelder, 1819.

10. Weslager, pp. 86–97.

11. E.W. Voegelin, 1940.

12. Newcombe, p. 58.

13. *Walam Olum or Red Score. The Migration Legend of the Lenni Lenape or Delaware Indians,* Indiana Historical Society, Indianapolis, 1954. The quotes by Black are on p. 333.

14. Oestreicher, 1994, and 1996.

15. Oestreicher, 1994, pp. 19–20; 1996, p. 12; see Boewe, 1987; Oestreicher quotes from a 1988 preprint of Boewe.

16. Rafinesque, "The Americans are not Jews," *Saturday Evening Post,* vol. 8, 1829, p. 12; *Atlantic Journal and Friends, of Knowledge,* vol. 1, 1832, pp. 98–99.

17. Williams, 1991, p. 108. For an informative discussion of fraud in science see Sapp, 1990, pp. 1–26.

Chapter 13. Botany and Zoology

1. Clay, 1942. In his Kentucky explorations, Rafinesque named the copperhead snake (*Agkistrodon mokasen cupreus),* the mudpuppy (*Necturus maculosus*), the cave salamander (*Eurycea lucifuga*), and two other salamanders, an achievement few if any specialists in snakes and reptiles could match. Among mammals, he identified and named certain bats, moles, jaguars, cougars, deer, and gophers.

2. Jordan, 1886, p. 220.

3. All four works have been translated into English, edited, and discussed by Cain, 1990.

4. Cain, p. 62.

5. The monograph is owned by the Academy of Natural Sciences of Philadelphia. I thank Earle E. Spamer for pointing out the uncut state of the copy.

6. Corti, pp. 1–17.

7. Allen, 1976.

8. Some of Rafinesque's specimens can be found in the Muséum National d'Histoire Naturelle (Jardin des Plantes), Paris, the Academy of Natural Sciences of Philadelphia, and the Darlington Herbarium at West Chester, Pa.

9. Rafinesque to Z. Collins, August 12, 1818; see Pennell, 1942, Stuckey, 1998, pp. 134–35. On his trip down the Ohio River, he not only worked on fish and molluscs, he collected six hundred species of plants of which twenty were new.

10. Rafinesque, *Kentucky Gazette,* April 4, 1822; Stuckey, 1998, pp. 134–35, 139.

11. Boewe, 1983, pp. 10–11. A lecture on botany by Rafinesque, edited by C. Boewe.

12. Boewe, cited in his *On Botany*, 1983.

13. Stuckey, 1998, p. 113.

14. Merrill, 1949, p. 7.

15. Ibid,. 1943, p. 113.

16. Friesner, 1953.

17. Gray, 1841, discussed Rafinesque's work published in his *Neogenyton,* 1826.

18. Merrill, 1942, p. 85.

19. Pennell has published eight papers on Rafinesque's overlooked names, while Merrill has contributed eleven. Merrill's *Index Rafinesquianus* is a definitive account of Rafinesque's taxon-creating proclivities and a superb summary of Rafinesque's life and work. See also Stuckey, 1971b.

20. Britton, 1888.

21. Stuckey, 1998, p. 125; the eighth edition of Gray's manual was edited in 1950 by Merritt, L. Fernald.

22. Stuckey, 1997 (Kentucky), 1998, (Ohio River Valley).

23. Swainson, pp. 300–301.

24. The two major publications are *Carattei di alcuni nuova generi e nuove specie di animali e animali* and *Indice d'ittiologia Siciliana.* Swainson supervised the printing of the *Indice* at Messina.

25. Haldeman, 1842. A complete list of Rafinesque's writings on fish was published by Dean, 1917.

26. Rafinesque published observations on forty fish of the northeast and a memoir on Sturgeon.

27. Markle, 1997.

28. Agassiz, 1854.

29. Quoted in Keir B. Sterling's Introduction to Rafinesque's autobiography, published by Arno Press Inc.,N.Y., 1976, p. xii.

30. Girard, C. *Proceedings of the Academy of Natural Sciences of Philadelphia Journal,* vol. 8, 1856, pp. 165–68; Agassiz, L *American Journal of Science,* series 2, vol. 19, pp. 71–99, pp. 215–31; also see Jordan, 1877.

31. Copeland, 1876.

32. Jordan, 1877, pp. 5–53.

33. Copeland, pp. 472–73.

34. Call, p. 96.

35. Rafinesque, *American Monthly Magazine and Critical Review*, vol. 1, 1817, pp. 431–35. A paper with the same title was published in the *Philosophical Magazine and Journal*, vol. 54, 1819, pp. 361–67.

36. Stroud, 1992.

37. Lesueur, C.A., *Proceedings of the Academy of Natural Sciences of Philadelphia Journal,* vol. 1, May 1817 (reporting on Mediterranean molluscs); Say, Thomas, Ten papers in the *Proceedings of the Academy of Natural Sciences of Philadelphia Journal,* vols. 1, 2, 4, 5 (1817–1826); Rafinesque, *American Monthly and Critical Review,* vols. 3, 4 (1818–1819); *Journal de Physique*, Paris, June, 1819. A short history of Conchology in the United States was written by George W. Tryon, in *The American Journal of Science and Arts*, (second series), vol. 33, 1862, pp. 161–80.

38. Bogan, 1988, has written an excellent review of Rafinesque's work on freshwater bivalves.

39. Johnson, 1973.

40. Rafinesque, *Journal de Physique*, vol. 88, 1819, p. 423. Theodore Gill brought attention to this ignored paper meriting priority in *Proceedings of the Academy of Natural Sciences of Philadelphia Journal,* 1864, p. 152.

41. Lea published 13 volumes on molluscs between 1829 and 1872.

42. Conrad, 1853. See Lea, for a detailed rebuttal of Conrad, in *Proceedings of the Academy of Natural Sciences of Philadelphia Journal,* vol. 7, 1854, pp. 236–49.

43. Rafinesque in *Annals of Nature,* 1820, p. 71: "Since 1820, several American Conchologists have attempted to notice, describe, or figure these shells; Barnes, in 1823, Lea, Say, and Eaton, later still. They had a fine field before them, in elucidating them by good figures, and describing the new kinds; but led astray, by various motives, they have neglected to verify, or properly notice my previous labors, *although they were known to them.* Mr. Say is above all, inexcusable. I had respectfully noticed, in 1820, his previous labors; but he has never mentioned mine, and knows so little of the animals of these shells, as to have mistaken their mouth for their tail, and their anterior for the posterior part of the shells! . . . "I ought to have added to the names of our late writers on *Unio,* Mr. Hildreth, who has described over again a few of my species, and Prof. Eaton, who I regret to say, has, (in his Zoological Textbook, Albany, 1826, now before me,) noticed 33 species of *Unio* and *Elasmodon* of Say and Barnes, but none of my previous ones! and put them all back to the old genus *Mya* of Linnaeus! This, as his whole zoological book, proves that he is forty years backwards in the science of Zoology, as he is 30 years backwards in Botany, and about 20 in Geology. But this is not peculiar to him, it is the fate of one half of our Naturalists, Botanists and Geologists. The daily increase in of knowledge and improvement in science is de-

spised or neglected by them as useless innovation! While all the world, and all the sciences move forward, they would keep those they teach or cultivate at a stand! it is all in vain, and time will show it." See also *The Monthly American Journal of Geology and Natural Sciences,* vol. 1, pp. 370–77, 1831. The translation was done by C.A. Poulson.

44. Binney, William G. in his father Amos's treatise on air-breathing molluscs, pp. 38–39.

45. Lea, 1870, Introduction. (Rafinesque became a citizen in 1832.)

46. Jordan, Geological Survey of Ohio, vol. 4, 1882, p. 741, quoted in Call, p. 98.

47. Walker, B.,The Rafinesque-Poulson Unios, *Nautilus,* vol. 30, 1916, pp. 43–47.

48. Call, p. 96–102.

49. Ortmann, pp. 335–38.

50. Rhoads, 1911.

51. Ibid., pp. 6–7.

52. Richmond, 1909, p. 38. A detailed account of Rafinesque's ornithological work can be found in Rhoads, 1911 (published by the Delaware Valley Ornithological Club).

53. Rhoads reprinted two of Rafinesque's *Kentucky Gazette* articles (Feb. 14 and Feb. 21, 1822) in *The Auk,* vol. 29, 1912, pp. 191–98.

54. Porter, 1986, pp. 41–51.

55. G. Ord to C. Waterton, April 23, 1832, (APS).

56. Rafinesque, *Kentucky Gazette,* vol. 1, 1822, p. 3; Ibid.

57. Holthuis, 1954.

Chapter 14. Last Years in Philadelphia

1. Wainwright in *Philadelphia, A 300-year History,* ed. R.F. Weigley, pp. 258–306; Repplier, 1898. DeKay (1826) summarizing the state of American science in 1825 and stressing the need for improving scientific education gives some significant information about libraries and books in four major U. S. cities: New York, with a population of 170,000 had ten public libraries containing 44,000 books. The corresponding numbers for Baltimore were 70,000 / 4 / 30,000; for Philadelphia, 160,000 / 19 / 70,000; for Boston, 60,000 / 13 / 55,000, (p. 74). New York had overtaken Philadelphia in size by 1810.

2. Much of this chapter and many of the quotes are taken from Rafinesque's autobiographical *A Life of Travels.*

3. Ten notebooks are in the possession of the Smithsonian Institution. An account of Rafinesque's notebooks has been written by Boewe, Reynaud, and Seaton in their introduction to the French version of a *Life of Travels.*

4. See Boewe et al., Introduction to the French version of a *Life of Travels,* p. 13, 1987.

5. The quote is from a volume written by Edwin C. Jellett who wrote *Recollections of William C. Kite.* Kite, a student, attended Rafinesque's lectures. Jellett's account is found in Smyth, pp. 326–27.

6. Rafinesque's finances, based on his *Day-Book,* a manuscript at the Library Company, are discussed by Pennell, pp. 41–43, and Smyth, pp. 331–34.

7. Rafinesque, *Atlantic Journal and Friend of Knowledge,* 1833, pp. 200–201.

8. See Pennell, pp. 43–45.

9. See Montgomery, which contains the Wagner-Rafinesque correspondence, held by the Wagner Free Institute of Science of Philadelphia.

10. Stone, pp. 170–71, quotes a letter dated October 25, 1825, from a Dr. Edmund Porter of Frenchtown, N.J., to Dr. Thomas Miner of Haddam, Conn.

11. Featherstonhaugh, G.W., *The Monthly American Journal of Geology and Natural Sciences,* vol. 1, 1832, pp. 508–15.

12. The two Rafinesque works referred to are a pamphlet entitled " Enumeration and account of some remarkable natural objects of the Cabinet of Professor Rafinesque in Philadelphia, being Animals, Shells Plants, and Fossils, collected by him in North America between 1816 and 1831"; and *Atlantic Journal and Friend of Knowledge, a Cyclopedic Journal and Review of Universal Science and Knowledge . . . ,* Editor C. S. Rafinesque.

13. Rafinesque, *Atlantic Journal and Friend of Knowledge,* vol. 1, 1832, pp. 110–14; the *Rhinoceroides* paper is on pp. 114–15 of the same journal, vol. 1, 1832. Two decades later, Joseph Leidy described teeth belonging to the rhinoceros—definitive evidence for the presence of this beast in North America. However Graustein (p. 108) states that John D. Clifford had collected teeth of the rhinoceros "from Big Bone Lick and similar sites in Kentucky." These were in his cabinet in 1820.

14. Richard Harlan, *Fauna Americana,* 1825, Introduction, p. viii; see pp. 302–9.

15. Seaton, 1988.

16. Rafinesque, "Description d'une Ville Ancienne du Kentucky occidental sur la Rivière Cumberland," *Bulletin de la Société de Geographie,* vol. 20, 1833, pp. 236–41.

17. Ibid., *Atlantic Journal,* vol. 1, 1835, pp. 187–88.

18. See Novack, 1967, 1980, for an insightful discussion of eighteenth-century art and science.

19. Letters from Rafinesque's sister Georgette, and his daughter, dating from 1830 to 1834 are quoted extensively by Pennell, pp. 45–53. Letters from Rafinesque to his sister or daughter have not been found. The orphan school for girls would be modeled after the school for white male orphans founded by the Philadelphia banker Stephen Girard.

20. Haldeman, 1842; Haldeman left the Rafinesque material to the ANSP which later went to the APS; a discussion of the subject is found in the Introduction to a reprinting of the original version of *A Life of Travels* (in French), by C. Boewe, G. Raynaud, and Beverly Seaton.

21. Unfortunately Charles Wetherill died in 1838, two years before Rafinesque.

22. Paclt, J., 1960.

23. Rafinesque, see Introductions to the four parts of *Flora Telluriana.*

24. Gray, 1841, pp. 238–39.

Chapter 15. Last Days

1. Rafinesque to Torrey, October 7, 1836.

2. Description by Joseph Henry dated May 29, 1826, cited in Reingold, 1972, vol. 1, 1972, pp. 152–53.

3. Pennell, 1942, pp. 61–62 (footnote).

4. Call, pp. 55–56.

5. See Lippard, 1876.

6. H. H. reprinted from the *Philadelphia Ledger Supplement, May 5, 1877,* in the *American Naturalist,* vol. 11, 1877, pp. 574–75.

7. Meehan, Thomas, *Botanical Gazette*, vol. 8, 1883, pp. 177–78.

8. Pennell, 1950. See also Pennell, 1940; Pennell, 1942, pp. 62–67; Hallowell, 1840.

9. Boewe, 1987b. Many of the details in this chapter have been taken from this fascinating paper.

10. Hallowell, 1840.

11. Rafinesque to Swainson, March 15, April 10, April 12, 1840 (Linnean Society Archives).

12. Boewe, vol. 18, 1999, p. 69.

13. James Ronaldson, owner of a print foundry, was the first president of the Franklin Institute. According to Hance, (pp. 528–29) he probably knew Rafinesque, and his type was probably used for Rafinesque's Philadelphia publications.

14. In earlier reports such as those of Smyth (p. 339) and Fitzpatrick (p 42), the cemetery was located at Ninth Street and Catharine Street, about three blocks away from Hance's and Boewe's location at the southwest corner of Ninth and Bainbridge Streets, which is today a playground. (See Hance, p. 529). Hance incorrectly states that Rafinesque was sixty-three years of age when he died—he was fifty-seven.

15. Merrill, 1949, pp. 33–37; Pennell, 1945; Stuckey, 1971; Chase, 1936. Some of his specimens (approximately 275) ended up in the Academy of Natural Sciences of Philadelphia, and a few others at Harvard University, the New York Botanical Garden, and at the Darlington Herbarium of West Chester State University, Pennsylvania. His salvaged manuscripts can be found scattered about the country—at the Academy of Natural Sciences, the American Philosophical Society, and the Library Company in Philadelphia, Harvard University, the Smithsonian Institute in Washington, Transylvania University in Lexington, Kentucky, and at the University of Kansas. In his papers and in his 1943 publication, Merrill, a history-minded botanist who was one of the major champions of Rafinesque, discusses the basis for the turmoil Rafinesque caused in American and world botany. The present discussion is based on these publications and that of Pennell, 1942.

16. Details in the will are given in Call, and are discussed by Smyth, pp. 340–41.

17. The medal is now in the possession of the College of Physicians of Philadelphia.

18. During the Yellow Fever epidemic of the early 1790s, Girard founded a hospital and personally tended the afflicted until the scourge subsided. He founded a school for male white orphans, housed in a building that was considered by some to be the finest building in the United States. When he died, he left his entire fortune to the city of Philadelphia.

19. Hance, 1917.

20. A different version of this story is found in a short letter in *Science,* vol. 59, 1924, pp. 553–54, by David Starr Jordan. According to Jordan, Mrs. Norton had heard of Mercer's memorial tablet and asked for a photograph of the stone, which she would present to the Rafinesque Botanical Club. Later, upon learning that the cemetery was to be destroyed, she enlisted her brother, James A. Spencer of Philadelphia, to urge appropriate officials to allow the transfer of Rafinesque's remains to Lexington.

21. Boewe, 1987b; Wright, pp. 353–56.

22. Jordan, 1927.

23. Boewe, 1987b, pp. 219–21.

Epilogue

1. Ewan, 1975.

2. David Starr Jordan (1888) states that the ornithologist Elliot Coues made the suggestion (p. 146). The word *rafinesque* would be distinct from the harsh and derogatory *raffish* which according to Webster's Dictionary (2nd edition) means (1) disreputable; disgraceful, and (2) tawdry; flash; cheap. These words do not really apply to Rafinesque, who was an honorable, idealistic victim of his imagination, enthusiastic, grandiose in his visions, part charlatan, and forever bringing calamity down upon himself.

3. Quatre, 1948.

4. Rafinesque, *Western Review and Miscellaneous Magazine*, vol. 1, 1819, pp. 60–62.

5. Gray, 1841.

6. Comments were collected by R. L. Stuckey and published in 1986.

7. Cain, 1990, p. 23.

8. Lurie, 1988, chapter 4, *The American Welcome, 1846–1850,* pp. 122–65. Agassiz statement about Rafinesque is from the *American Journal of Science and Arts,* 1854, p. 354, and is quoted in Starr, 1888, p. 156.

9. Rafinesque, On the Salivation of Horses, *Western Review and Miscellaneous Magazine*, vol. 1, 1819, pp. 182–84.

10. Ibid., *American Monthly Magazine and Critical Review*, vol. 3, 1818, pp. 41.

11. Merrill, 1949, pp. 54–56.

12. Pratt and Boewe, 1992.

13. The list comes from *Narcissistic Personality Disorder*, in *Diagnostic and statistical manual of mental disorders,* 4th edition, American Psychiatric Association, Washington, D.C., 1994. For excellent, relevant discussions of manic depression, see Jamison, who examines the question of manic depressive mood disorders and artistic creativity.

14. Winchester, 1998.

15. Letter of October 3, 1859 (Gray Herbarium Library, Harvard).

16. See Sapp, pp. 1–26 for a discussion of scientific dishonesty.

17. Rafinesque, *Ichthyologia Ohiensis,* 1820.

BIBLIOGRAPHY

Allen, D. E., *Natural History and Social History,* Journal of the Society for the Bibliography of Natural History, vol. 7, 1965, pp. 509–16.

Atwater, Caleb, *Description of the Antiquities Discovered in the State of Ohio and Other Western States* in Transactions and Collections of the American Antiquarian Society, vol. 1, 1820.

Audubon, John James, *The Eccentric Naturalist* in *Ornithological Biography,* Edinburgh, 1831.

———. *Delineations of American Scenery and Character,* Introduction by F.H. Hobart; Simkin, Marshall, Hamilton, Kent & Co., LTD, London, 1926.

Baatz, Simon, *Patronage, Science and Ideology in an American City: Patrician Philadelphia, 1800–1860,* Ph.D. Thesis, University of Pennsylvania, 1986.

———. *Knowledge, Culture, and Science in the Metropolis, The New York Academy of Sciences, 1817–1970,* New York Academy of Sciences, N.Y., 1990.

Barkley, A. H., *Constantine Samuel Rafinesque,* Annals of Medical History, (series 1), vol. 10, 1928, pp. 66–76.

Barlow, William, and Powell, David O., "The Late Dr. Ward of Indiana: Rafinesque's Source of the *Walam Olum,*" *Indiana Magazine of History,* vol. 82, 1986, pp. 185–93.

Barnhart, John H., *Some American Botanists of Former Days,* Journal of the New York Botanical Garden, vol. 10, 1909, pp. 177–90.

Barton, Benjamin Smith, *New Views of the Origin of Tribes and Nations of America,* 2nd ed, Philadelphia, 1798.

Belyi, Vilen B., *Rafinesque's Linguistic Activity,* Anthropological Linguistics, vol. 39, 1997, pp. 60–73.

Berman, Alex, *C.S, Rafinesque (1783–1840): a Challenge to the Historian of Pharmacy,* American Journal of Pharmaceutical Education, vol. 16, 1952, pp. 409–18.

———. *A Striving for Scientific Respectability: American Botanics and the Nineteenth Century Plant Materia Medica,* Bulletin of the History of Medicine, vol. 30, 1956, pp. 7–31.

———. *An Unpublished Letter from G. K. Lawrence to C. S. Rafinesque, October 8, 1828,* Bulletin of the History of Medicine, vol. 34, 1960, pp. 461–70.

Betts, Edwin M., *The Correspondence Between Constantine Samuel Rafinesque and Thomas Jefferson,* Proc. American Philosophical Society, vol. 87, 1944, pp. 368–80.

Binney, Amos,*Terrestial Air-Breathing Mollusks of the United States*, ed. A.A. Gould, vol. 1, 1851.

Blake, William, *Songs of Innocence and of Experience,* ed. Ruth E. Everett, Avon Books, N.Y., 1971.

Boewe, Charles, *Rafinesque and Dr. Short,* Filson Club Quarterly, vol. 35, 1961, pp. 28–32.

———. *Mt. Rafinesque,* Names, vol. 10, 1962, pp. 58–60.

———. *Fitzpatrick's Rafinesque: A Sketch of his Life With Bibliography,* Revised and Enlarged by Charles Boewe, M&S Press, Weston, Mass., 1982.

———. *C.S. Rafinesque on Botany (1820),* Edited with an Introduction and Notes, The Whippoorwill Press, Frankfort, Ky., 1983.

———. *A Note on Rafinesque, the Walum Olum, The Book of Mormon, and the Mayan Glyphs,* Numen, vol. 32, Fasc 1, 1985, pp. 101–13.

———. *The Fall From Grace of that "Base Wretch" Rafinesque,* Kentucky Review, vol. 7, 1987, pp. 39–53.

———. *The Walam Olum and Dr. Ward, Again,* Indiana Magazine of History, vol. 83, 1987a, pp. 344–59.

———. "Who's Buried in Rafinesque's Tomb?" *Pennsylvania Magazine of History and Biography,* vol. 111, 1987b, pp. 213–35.

———. *Introduction of Tea Plant at Bartram's Garden,* Bartram Broadside, Fall, 1994.

———. *Rafinesque, Constantine Samuel,* New Oxford Biographical Dictionary, vol. 18, 1999, pp. 68–70.

———. John D. Clifford's., *Indian Antiquities, related materials by C. S. Rafinesque,* University of Tennessee Press, Knoxville, 2000.

Bogan, Arthur, E., *A Bibliographic History of C.S. Rafinesque's Work on North American Freshwater Bivalves,* Archives of Natural History, vol. 15, 1988, pp. 149–54.

Bomare, Valmont de, *Dictionnaire Raisonné Universel D'Histoire Naturelle contenant L'Histoire des Animaux, des Végétaux et des Minéraux,* 6 vols. 1768, Paris.

Bourbourg, Brasseur de, *Quatre Lettres sur Le Mexique,* Paris, 1868.

Brinton, Daniel G., *The Lenape and Their Legends with the Complete Text and Symbols of the WALAM OLUM,* Philadelphia, 1885.

Britton, N. L., *The Genus Hicoria of Rafinesque,* Bull of the Torrey Botanical Club, vol. 15, 1888, pp. 277–85.

Britton, N.L., and Brown, A., *An Illustrated Flora of the Northern United States, Canada and the British Possessions,* Charles Scribner's Sons, 3 vols., 2nd ed., N.Y., 1913.

Bronowski, J., *William Blake, 1757–1827, A Man Without a Mask,* Secker and Warburg, London, 1947.

Bruce, Robert V., *The Launching of Modern American Science, 1846–1876,* Cornell University Press, Ithaca, N.Y., 1988.

Cain, Arthur J., *Constantine Samuel Schmaltz on Classification, a Translation of Early Works by Rafinesque with Introduction and Notes,* Tryonia No. 20, Academy of Natural Sciences of Philadelphia, 1990.

Caldwell, Charles, *A Discourse on the Genius and Character of the Reverend Horace Holley, Ll.D.* Boston, Mass., 1828.

Call, Richard Ellsworth, *The Life and Writings of Rafinesque,* John P. Morton, Printers to the Filson Club, Louisville Ky., 1895.

Chase, Agnes, *The Durand Herbarium,* Bartonia, vol. 17, 1936, pp. 40–45.

Cheyney, Edward P., *History of the University of Pennsylvania, 1740–1940,* University of Pennsylvania Press, Philadelphia, 1940.

Clay, William M., *Herpetology and Rafinesque,* in Rafinesque Memorial Papers, October 31, 1940, Transylvania College Bulletin, vol. 15, no. 7, 1942, pp. 84–91.

Conrad, T.A., reporting on *A Synopsis of the Family of Naiades of North America,* by Isaac Lea, Proceedings of the Academy of Natural Sciences of Philadelphia Journal, vol. 6, 1853, pp. 243–69.

Copeland, Herbert E. *A Neglected Naturalist,* American Naturalist, vol. 10, 1876, pp. 469–73.

Corsi, Pietro, *The Age of Lamarck, Evolutionary Theories in France, 1790–1830,* University of California Press, Berkeley, 1988.

Curti, M., Shryock, R.H., Cochran, T.C., and Harrington, F. H., *An American History,* Harper & Bros., New York, vol. 1, 1950.

Daniels, George H., *The Process of Professionalization in American Science—the emergent period,* (1820–1860) Isis, vol. 58, 1967, pp. 151–66.

———. *American Science in the Age of Jackson,* Columbia University Press, New York, 1968.

———, ed., *Nineteenth-Century American Science: a Reappraisal,* Northwestern University Press, 1972.

Darwin, Charles., *On the Origin of Species by Means of Natural Selection,* 6th and last edition, Appleton & Co., N.Y., 1901.

———. *The Foundations of the Origin of Species, a Sketch Written in 1842,* ed by Francis Darwin, Cambridge University Press, 1909.

Dean, Bashford, *A Bibliography of Fishes,* vol. 2, ed. C.R. Eastman, American Museum of Natural History, 1917.

De Kay, James E., *Anniversary Address on the Progress of the Natural Sciences in the United States,* New York, 1826.

Dickens, Charles, *American Notes For General Circulation,* 1842, reprinted by Penguin Books, 1972.

———. *The Life and Adventures of Martin Chuzzlewit,* Oxford, 1966; first published in serial form in 1843.

Doskey, John S., *The European Journals of William Maclure,* American Philosophical Society, 1988

Drake, Daniel, *Reviews and Bibliographical Notices,* The Western Journal of the Medical and Physical Sciences, vol. 3, 1830, pp. 393–420.

Dupre, Huntley, *Transylvania University and Rafinesque, 1819–1826,* Filson Club History Quarterly, vol. 35, 1961, pp. 110–21.

Ellis, Richard, *Monsters of the Sea, The history, natural history and mythology of the oceans' most fantastic creatures,* Doubleday, New York, 1991.

Ewan, J., *Constantine Samuel Rafinesque,* in Dictionary of Scientific Biography, ed C.C. Gillespie, C. Scribner and Sons, vol. 11, 1975, pp. 262–64.

Fitzpatrick, T.J., *Rafinesque, A Sketch of His Life with Bibliography,* The Historical Department of Iowa, Des Moines, Iowa, 1911.

Friesner, Ray C., *Rafinesque and the Taxonomy of Indiana Vascular Plants,* Butler University Botanical Studies, vol. 11, 1953, pp. 1–4.

Fulling, Edmund H., *Thomas Jefferson, His Interest in Plant Life as Revealed in his Writings II,* Bulletin of the Torrey Botanical Club, vol. 72, 1945, pp. 248–70.

Funkhouser, W.D., *An Eccentiric Egocentrist,* (privately printed), Lexington, Ky., 1940, pp. 23. In the Herbarium Library, Harvard University.

Gerbi, Antonello, *The Dispute of the New World, The History of a Polemic, 750–1900,* Translated by Jeremy Moyle, University of Pittsburgh Press, Pittsburgh, Pa., 1955.

Gillispie, Charles C., *Genesis and Geology,* Harper Torchbook, N.Y., 1959.

Gobar, A. and Hamon, J. H., *A Lamp in the Forest, Natural Philosophy in Transylvania University, 1799–1859,* Transylvania University Press, Lexington, Ky., 1982.

Graustein, Jeannette, *Thomas Nuttall, Naturalist,* Harvard University Press, Cambridge, Mass., 1967.

Gray, Asa, *Notice of the Botanical Writings of the Late C.S. Rafinesque,* American Journal of Science and the Arts, (1st series), vol. 40, 1841, pp. 221–41.

Greene, John C. *American Science in the Age of Jefferson,* Iowa State University Press, Ames, IA, 1984.

Haldeman, Samuel S., *Notice of the Zoological Writings of the Late C.S. Rafinesque,* American Journal of Science and Arts, vol. 42, 1842, pp. 280–91.

Hallowell, E., *Case of Cancer of the Stomach and Liver,* Medical Examiner, Sept. 19, 1840, pp. 597–99.

Hance, Anthony M., *Grave of Rafinesque, the Great Naturalist,* Bucks County Historical Society Papers, vol. 4, 1917, pp. 510–29.

Harlan, Richard, *Fauna Americana,* Philadelphia, 1825.

Haven, Samuel F., *Archaeology of the United States or Sketches, Historical and Bibliographical, of the Progress of Information and Opinion Respecting Vestiges of Antiquity in the United States,* Smithsonian Contribution to Knowledge, Washington, D.C., 1856.

Heckewelder, John, *An Account of the History, Manners and Customs of the Indian Nations,* Transactions of the Historical and Literary Committee of the American Philosophical Society, vol. 1, 1819, pp. 1–348.

Herrick, F. H., *Audubon the Naturalist, A History of His Life and Time,* vol. 1 (of 2), Appleton and Co., New York, 1917.

Hoffman, Daniel, *Brotherly Love,* Vintage Books, New York, 1981.

Holthuis, L.B., *C.S. Rafinesque as a Carcinologist, An Annotated Compilation of the Information on Crustacea Contained in the Works of that Author,* Zoologische Verhandelingen, E.J. Brill, Leiden, 1954.

Hopkins, J. F., ed *The Papers of Henry Clay,* University of Kentucky Press, vol. 4, 1959.

Jaffe, Bernard, *Men of Science in America: the Role of Science in the Growth of Our Country,* Simon and Schuster, New York, 1944.

Jamison, Kay Redfield, *Touched with Fire,* Free Press Paperbacks, New York, N.Y., 1993.

Jefferson, Thomas, *Notes on the State of Virginia,* ed. William Peden, Chapel Hill, N.C., 1781

Jennings, W.W., *Transylvania, Pioneer University of the West,* Pageant Press, New York, 1955.

Johns, Elizabeth, *Science, Art, and Literature in Federal America, Their Prospects in Federal America,* in *Everyday Life in the Early Republic,* ed. Catherine Hutchins, Winterthur Museum, Winterthur, Del., 1994.

Johnson, Paul, *The Birth of the Modern: World Society, 1815–1830,* Harper Collins, 1991.

Johnson, Richard I., *The Types of Unionidae (Mollusca: bivalva) described by C.S. Rafinesque in the Museum National D'Histoire Naturelle de Paris,* Journal de Conchyliologie, vol. 110, 1973, pp. 35–37.

Jordan, David Starr, *Review of Rafinesque Memoirs on North American Fishes,* in Contributions to North American Ichthyology, U.S. National Museum, Washington, 1877.

———. *A Sketch of Constantine Samuel Rafinesque,* The Popular Science Monthly, vol. 29, 1886, pp. 212–21.

———. *Science Sketches,* A.C. McClurg and Co., Chicago, 1888, pp. 143–59.

———. *The Bones of Rafinesque,* Science, vol. 59, 1924, pp. 554–55.

Jordon, David Starr, and Butler Amos W., *New Harmony,* Proceedings of the Indiana Academy of Science, vol. 37, 1927, pp. 59–62.

Keeney, Elizabeth B., *The Botanizers, Amateur Scientists in Nineteenth-Century America,* University of North Carolina Press, Chapel Hill, N.C.,1992.

Kohlstedt, Sally Gregory, ed,*The Origins of Natural Science in America, The essays of George Browne Goode,* Smithsonian Institution Press, Washington, D.C., 1991.

Lea, Isaac, *Observations on the Genus Unio, Together with Descriptions of New Genera and Species,* 1832.

———. *Synopsis of the Family, Unionidae,* 14th ed., 1870.

Lee, Rebecca Smith, *Mary Austin Holley, A Biography,* University of Texas Press, Austin, Texas, 1962.

Lindley, John, *An Introduction to the Natural System of Botany,* First American Edition, G., C., and H. Carvill, New York, 1831.

Lippard, George,*The Quaker City; or, The Monk of Monk Hall, A Romance of Philadelphia Life, Mystery and Crime,* Peterson, Philadelphia, 1876.

Little, Elbert L., Jr., *A Note on Rafinesque's Florula Columbica,* Proceeds of the Biological Society of Washington, vol. 56, 1943, pp. 57–66.

Lurie, Edward, *Louis Agassiz, A Life in Science,* Johns Hopkins University Press, Baltimore, Md., 1988. (Originally published in 1960 by the University of Chicago Press).

Maclure, William. Observations on the Geology of the United States of America, Philadelphia, Pa., 1817.

Markle, Douglas F., *Audubon's hoax: Ohio River Fishes described by Rafinesque,* Archives of Natural History, vol. 24, 1997, pp. 439–47.

Mayr, E. *Populations, Species and Evolution,* Belknap Press of Harvard University Press, 1977.

McAllister, Ethel M., *Amos Eaton, Scientist and Educator, 1776–1842,* University of Pennsylvania Press, Philadelphia, 1941.

McCulloh, J.H., *Researches, Philosophical and Antiquarian, Concerning the Aboriginal History of America,* 1829.

McCutcheon, David, *the Red Record, the Wallam Olum, the Oldest Native North American History,* Avery Publishing Group, Garden City Park, N.Y., 1993.

Merrill, Elmer D. *A Generally Overlooked Rafinesque Paper,* Proceedings of the American Philosophical Society, vol. 86, 1942, pp. 72–90.

———. *Rafinesque's Publications from the Standpoint of World Botany,* Proceedings of the American Philosophical Society, vol. 87, 1943, pp. 110–19.

———. *C.S. Rafinesque,with Notes on his Publications in the Harvard Libraries,* Harvard Library Bulletin, vol. 2, 1948, pp. 5–21.

———. *Index Rafinesquianus,* Jamaica Plain, Mass., Arnold Arboretum, 1949.

Miller, Amy Bess, *Shaker Herbs, A History and a Compendium,* Clarkson N. Potter Inc., New York, 1976.
Montgomery,T. L., *Correspondence of C.S. Rafinesque and Professor Wm. Wagner,* Science, Vol. 11, 1900, pp. 449–51.
Morton, Samuel George, *A Memoire of William Maclure,* Proceedings of the Academy of Natural Sciences of Philadelphia Journal, vol. 1, 1841;
Myer, Clarence, *American Folk Medicine,* Meyerbook Publisher, Glenwood, Ill., 1973.
Newcombe, William W. Jr., *The Walam Olum of the Delaware Indians in Perspective,* The Texas Journal of Science, vol. 7, 1955, pp. 57–63.
Novak, Barbara, *American Painting of the Nineteenth Century: Realism, Idealism and the American Experience,* Praeger, New York, 1969.
———. *Nature and Culture: American Landscape and Painting,* 1825–1875, Thames and Hudson, London, 1980.
Oestreicher, David M., *Unmasking the Walam Olum, a 19th Century Hoax,* Bulletin of the Archaeological Society of New Jersey, vol. 49, 1994, pp. 1–43.
———. *Unravelling the Walam Olum,* Natural History, October, 1996.
Ord, George, *A Memoir of Thomas Say,* (1834), in *The Complete Writings of Thomas Say on the Entomology of North America,* ed. John L. LeConte, vol. 1, 1859, pp. vii–xiv, Balliere Brothers, N.Y. Reprinted by Arno Press, New York, 1979.
Ortmann, A.E., *A Monograph of the Najades of Pennsylvania,* Memoire of the Carnegie Museum, vol. 4, 1911, pp. 279–343.
Paclt, J., *The "Classes" of Rafinesque and the Modern Biosystematics,* Taxon, vol. 9, 1960, pp. 47–49.
Peattie, Donald Culcross, *Green Laurels, The Lives and Achievements of the Great Naturalists,* Simon and Schuster, New York, 1936.
Pennell, Francis W., *New Light on Rafinesque,* Chronica Botanica, vol. 6, 1940, pp. 125–26.
———. *The Life and Work of Rafinesque,* in Rafinesque Memorial Papers , October 31, 1940, edited by L.A. Brown. Lexington: Transylvania College Bulletin, vol. 15, 1942.
———. *The Last Sickness of Rafinesque,* Chronica Botanica, vol. 12, 1950, pp. 216–17; Bartonia, no. 25, 1949, pp. 67–68.
Perkins, Samuel E. III, *Letters by Rafinesque to Dr. Short in the Filson Archives* , Filson Club History Quarterly, vol. 12, 1938, pp. 200–239.
Pierson, George Wilson, *Tocqueville in America,* Doubleday Anchor Book, Garden City, N.Y., 1959.
Pietsch, T. W., *Historical Portrait of the Progress of Ichthyology from its Origin to our Own Time, George Cuvier,* Johns Hopkins Press, Baltimore. Md., 1995.
Pluche, Abbé le, *Spectacle de la Nature or Nature Delineated being Philosophical Conversations,* 1760, Paris.
Porter, Charlotte M., *'Subsilentio' : discouraged works of early nineteenth-century American natural history,* Journal of the Society for the Bibliography of Natural History, vol. 9, 1979, pp. 109–19.
———. *The Concussion of Revolution: Publications and Reform at the Early Academy of Natural Sciences, Philadelphia, 1812–1842,* Journal of the History of Biology, vol. 12, 1979a, pp. 273–94.
———. *The Eagle's Nest, Natural History and American Ideas, 1812–1842,* University of Alabama Press, Tuscaloosa, Ala., 1986.
Priest, Josiah, *American Antiquities and Discoveries in the West,* Albany, N.Y., 1838.

Quatre, Henricus, *Rafinesque, A Concrete Case,* Archivio Botanico, 3rd series, vol. 24, 1948, pp. 2–18.

Rafinesque–Schmaltz, Constantine Samuel, *Précis des decouvertes et travaux somiologiques,* Palerme, 1814.

Rafinesque, C.S., *On the Different Lightnings Observed in the Western States,* Western Review and Miscellaneous Magazine, vol. 1, 1819, pp. 60–62.

———. *Ichthyologia Ohiensis, or Natural History of the Fishes Inhabiting the River Ohio,* Lexington, Ky., 1820.

———. *Ancient History or Annals of Kentucky,* Frankfort, Ky., 1824.

———. *The Pulmist; or Introduction to the art of Curing and Preventing The Consumption or Chronic Phthisis,* Philadelphia, Pa., 1829.

———. *Medical Flora, or Manual of the Medical Botany of the United States of North America,* Philadelphia, Pa., vol. 1, 1828; vol. 2, 1830.

———. *First Letter to Mr. Champollion, on the Graphic Systems of America, and the Glyphs of Otolum or Palenque, in Central America,* Atlantic Journal, vol. 1,1832.

———. *A Life of Travels and Researches in North America and South Europe or The Life, Travels and Researches of C.S.Rafinesque, A.M. Ph.D.* Philadelphia, Pa., 1836. Republished in Chronica Botanica, vol. 8, 1944, pp. 292–360. Foreward by Elmer D. Merrill and Critical Index by Francis W. Pennell. There is also an original French version with an introduction by C. Boewe, Georges Reynaud, and Beverly Seaton who also edited the work. *Précis ou Abrégé de Voyages, Travaux, et Recherches de C.S. Rafinesque,* North-Holland Publishing Company, Amsterdam, 1987.

———. *The American Nations; or Outlines of their General History, Ancient and Modern,* Philadelphia, Pa., 1836.

———. *New Flora of North America,* 1836, 1838.

———. *The World, or Instability, A Poem, with notes and illustrations,* Philadelphia Pa., 1836. Reprinted in 1956 by Scholars' Facsimiles and Reprints, Gainesville, Fla. Introduction by Charles Boewe.

———. *Flora Telluriana,* in 4 parts, 1836–37.

———. *Safe Banking including The Principles of Wealth,* Philadelphia, Pa., 1837.

———. *The Ancient Monuments of North and South America,* Philadelphia, Pa., 1838.

———. *The Good Book, Amenities of Nature or Annals of Historical and Natural Sciences,* number 1, Philadelphia, Pa., January, 1840.

———. *DayBook of C.S. Rafinesque, 1832 to 1834,* manuscript in the archives of the Library Company (Philadelphia).

———. *Walam Olum, or Red Score, The Migration Legend of the Lenni Lenape or Delaware Indians,* (A new translation and a collection of scholarly papers by experts on the subject) Indiana Historical Society, Indianapolis, Ind., 1954.

Reingold, N. ed. *The Papers of Joseph Henry,* vol. 1, Smithsonian Institution Press, Washington, D.C., 1972.

Rezneck, Samuel, *A Travelling School of Science on the Erie Canal in 1826,* New York History, vol. 40, 1959, pp. 255–69.

Reynaud, Georges, *Un grande naturaliste meconnu: Constantine Samuel Rafinesque (1783–1840).* [Revue municipale] Marseille, no. 112, 1978.

Rhoads, Samuel N., *Constantine S. Rafinesque as an Ornithologist,* Cassinia, no. 15, 1911, pp. 1–12.

———. *Additions to the Known Ornithological Publications of C.S. Rafinesque,* The Auk, vol. 24, 1912, pp. 191–98, (supplement on p. 401).

Richardson, Edgar P., *The Athens of America,* in *Philadelphia, A 300-year History,* W.W. Norton & Co., New York, 1982, pp. 208–57.

Richmond, Charles W., *A Reprint of the Ornithological Writings of C.S. Rafinesque,* The Auk, vol. 26, part 1, pp. 37–55, part 2, pp. 248–62, 1909.

Rodgers, Andrew D. III, *John Torrey, A Story of American Botany,* Princeton University Press, Princeton, N.J., 1942.

Rosenberg, Charles C. "The Book in the Sickroom: A Tradition of Print and Practice," in *Every Man His Own Doctor: Popular Medicine in Early America,* The Library Company of Philadelphia, 1998.

Rothstein, William J., *The Botanical Movements and Orthodox Medicine,* in Other Healers, Unorthodox Medicine in America, ed. Norman Gevitz, Johns Hopkins Press, Baltimore, Md., 1988.

Rusk, Ralph Leslie, *The Literature of the Middle Western Frontier,* (2 vols.), Columbia University Press, New York, 1925.

Saint-Pierre, Bernardin de, *Paul et Virginie,* translated by Raymond Hein, Moka, Mauritius: Editions de l'Océan Indien, 1981.

Sapp, Jan, *Where the Truth Lies, Franz Moewus and the Origins of Molecular Biology,* Cambridge University Press, Cambridge, Mass., 1990.

Seaton, Beverly, *Rafinesque's Sentimental Botany: The School of Flora,* Bartonia, no. 54, June 3, 1988, pp. 98–106.

Sheridan, Francis C., *Galveston Island, or a Few Months Off the Coast of Texas, (1839–1840),* edited by Willis W. Pratt. University of Texas Press, Austin, 1954.

Shinners, Lloyd H., *Polygonum bicorne Raf. Instead of P. longistylum Small,* Rhodora, vol. 59, pp. 265–67, 1957.

Shryock, Richard Harrison, *Medicine and Society in America: 1660–1860,* Cornell University Press, Ithaca, N.Y., 1960.

Simpson, George Gaylord, *Principle of Animal Taxonomy,* Columbia University Press, New York, 1961; reprinted 1990.

Slaughter, Thomas P., *The Natures of John and William Bartram,* Vintage Books, New York, 1996.

Smith, Billy G., ed., *Life in Early Philadelphia, Documents From the Revolutionary and Early National Periods,* The Pennsylvania State University Press, University Park, Pa., 1995.

Smyth, S. Gordon, *Rafinesque, The Errant Naturalist,* Historical Sketches, Historical Society of Montgomery County Pennsylvania, Norristown Press, vol. 6, pp. 300–342, 1929.

Squier, E. G. *Historical and Mythological Traditions of the Algonquins,* American Whig Review, Feb, 1849, pp. 273–93.

Stearn, W.T., *The Background of Linnaeus's Contribution to the Nomenclature and Methods of Systematic Biology,* Systematic Zoology, vol. 8, 1959, pp. 4–27.

Stevens, Peter F., *Mind, Memory and History: How classifications are shaped by and through time, and some consequences,* Zoologica Scripta, vol. 26, 1997, pp. 293–301.

Stewart, George E., *The Beginning of Maya Hieroglyphic Study: Contributions of Constantine S. Rafinesque and James H. McCulloh Jr.,* in Research Reports on Ancient Maya Writing no. 29, Center for Maya Research, Washington, D.C., 1989, pp. 11–28.

Stoltman, James B. *The Southeastern United States*, Chapter 5 in *The Development of North American Archaeology*, ed. James E. Fitting, The Pennsylvania State University Press, University Park, Pa., 1973, pp. 116–50.

Stone, Witmer, *Some Philadelphia Ornithological Collections and Collectors, 1784–1850,* The Auk, vol. 16, 1899, pp. 166–77.

Stroud, Patricia Tyson, *Thomas Say, New World Naturalist,* University of Pennsylvania Press, Philadelphia, Pa., 1992.

Stuckey, Ronald L., *The First Public Auction of an American Herbarium Including an Account of the Fate of the Baldwin, Collins and Rafinesque's Herbaria,* Talon, vol. 20, 1971a, pp. 443–59.

———. *C. S. Rafinesque's North American Vascular Plants at the Academy of Natural Sciences of Philadelphia,* Brittonia, vol. 23, 1971b, pp. 191–208.

———. *Opinions of Rafinesque Expressed by His American Contemporaries,* Bartonia, no. 52, 1986, pp. 26–41.

———. *Rafinesque's Botanical Pursuits in the Ohio Valley (1818–1826),* Journal of the Kentucky Academy of Science, vol. 59, 1998, pp. 11–157.

Stuckey, Ronald L., and Pringle, James S., *Common Names of vascular Plants reported by C. S. Rafinesque in an 1819 Descriptive Outline of four Vegetative Regions of Kentucky,* Transactions of the Kentucky Academy of Science, vol. 58, 1997, pp. 9–19.

Swainson, William, *Taxidermy, with the Biography of Zoologists*, vol. 119 of Lardner's Cyclopedia, London, 1840.

Thomas, Kieth, *Man and the Natural World,* Pantheon Books, New York, 1983.

Thomson, Samuel, *Family Botanic Medicine,* Boston, 1819.

———. *The Thomsonian Materia Medica, or Botanic Family Physician: Comprising a Medical Theory,* Albany, 12th ed., 1841.

Trollope, Frances, *Domestic Manners of the Americans*, 1832. Reprinted in 1951 by Alfred A. Knopf, New York.

Voegelin, E. W., *Cultural Parallels to the Delaware Walam Olum,* Procedures of the Indiana Academy of Science, vol. 49, 1940, pp. 28–31.

Wade, Richard C., *The Urban Frontier: Pioneer Life in Early Pittsburgh, Cincinnati, Lexington, Louisville, and St. Louis,* University of Chicago Press, Chicago, Ill., 1959.

Wainwright, Nicholas B., *The Age of Nicholas Biddle, 1825–1841,* in Philadelphia, A 300–Year History, ed Russel F. Weigley, W.W. Norton, New York, 1982.

Wallace, Daniel J., *Thomsonians: The People's Doctors,* Clio Medica, vol. 14, 1980, pp. 169–86.

Wauchope, Robert, *Lost Tribes and Sunken Continents, Myth and Method in the Study of American Indians,* University of Chicago Press, Chicago, Ill., 1962.

Webster, H. T., *Was Rafinesque an Eclectic?* California Medical Journal, vol. 19, 1898, pp. 344–45.

Weaks, Mabel Clare, *Medical Consultation on the Case of Daniel Vanslyke,* Bulletin of the History of Medicine, vol. 18, 1945, pp. 425–37.

Weiss, Harry B., *Rafinesque's Kentucky Friends*, Privately printed, Highland Park, N.J., 1936.

Weslager, C.A., *The Delaware Indians, A History,* Rutgers University Press, New Brunswick, N.J., 1972.

Williams, Stephen, *Fantastic Archaeology, The Wild Side of Archaeology, North American Prehistory,* University of Pennsylvania Press, Philadelphia, Pa., 1991.

Wilson, Alexander, *American Ornithology, or the Natural History of the Birds of the United States,* Philadelphia, Pa., 9 vols., 1808–1814.

Winchester, Simon, *The Professor and the Madman: a tale of murder, insanity, and the making of the Oxford English dictionary,* Harper Collins, 1998.

Wright, John D. Jr., *Transylvania: Tutor to the West,* University Press of Kentucky, Lexington, Ky., 1975.

INDEX

Abbreviation CSR refers to Constantine Samuel Rafinesque.